AF412726

Textbook of Anal Diseases

Textbook of Anal Diseases

Steen Lindkaer Jensen
*University of Århus
Denmark*

and

Ole Vagn Nielsen
*University of Copenhagen
Denmark*

harwood academic publishers
Australia • Canada • China • France • Germany • India • Japan
Luxembourg • Malaysia • The Netherlands • Russia • Singapore
Switzerland • Thailand • United Kingdom

British Library Cataloguing in Publication Data

Jensen, Steen Lindkaer
    Textbook of anal diseases
    1. Anus — Diseases
    I. Title   II. Nielsen, Ole Vagn
    616.3′5

ISBN 90-5702-299-0

Contents

Preface

After many years of work with patients with anal disease, we find that many of our colleagues are often insecure about these common diseases. Although many of these anal disorders are not considered as serious illnesses, they are nevertheless accompanied by a series of unpleasant symptoms which may be socially disadvantageous for the patient. Incorrect treatment may have lifelong consequences. It is important for all doctors to recognise the different anal disorders early and to institute correct treatment or at least not to start out on incorrect treatment. The method of treatment should be effective and simple and as gentle to the patient as possible. We have placed great importance on those forms of treatment that can be carried out as outpatient procedures.

The book is first and foremost written for junior doctors but also for those undergoing surgical training.

The book is based on our personal concepts and experiences gained over the years. It is not a systematic and detailed review of literature on the subject as the amount of literature is enormous, and there are many controversies. We have chosen to focus upon the methods of treatment which, from our own experience, can be safely and gently carried out but which at the same time help the patients.

It is our hope that the book will stimulate interest for this minor but quantitatively large area of gastroenterology for the benefit of the patients' quality of life. We have tried to make the book as realistic as possible by demonstrating the macroscopic and in many cases also the microscopic appearances of diseases by means of many colour photographs which have partly been made available by helpful and interested colleagues.

Steen Lindkaer Jensen
Ole Vagn Nielsen

Acknowledgements

The authors are grateful to the following colleagues for their generous contributions: Knud-Erik Sjølin, Jytte Roed-Petersen, Niels Rosman, Kaj Fischerman, Per Thommesen, Thorkil Christensen, Sten Mellerup Sørensen, Hans Gregersen and Lisbeth Jørgensen. All illustrations in the book have been drawn to our great satisfaction by Hanne Christiansen, medical illustrator at Rigshospitalet. We express our warmest thanks to all of them.

I. Anatomy and physiology of the anal region

The anus terminates the gastrointestinal tract and constitutes only a tiny part of it. In spite of this, normal structure and function of the anus is essential for maintaining continence of flatus and stools, and the organ is often the site of disorders that can have a detrimental influence on the patient's social and physical well-being. The anus is made up of the anal canal and an area of perianal skin.

The anal canal is a tube-like organ that histologically extends from the proximal to the distal edges of the internal anal sphincter (Fig. 1.1). Proximally it is in direct continuity with the rectum and distally the perianal skin. In formalin-fixed specimens the length of the anal canal is given as 3 cm on average. According to another definition, the anal canal extends from the anorectal ring to the anal opening and will thus be 4 cm long on average. The perianal skin is a diamond-shaped area that is longer antrolaterally than from side to side. It is bounded by the underside of the symphysis anteriorly, by the tip of the coccyx posteriorly, and by the genitofemoral fold and the underside of gluteus maximus laterally (Fig. 1.2). The anal canal is surrounded by powerful muscles, which by tonic contraction converts the canal to a collapsed cleft. The muscles around the anal canal can be regarded as two tubes, one surrounding the other (Fig. 1.1). The innermost visceral tube consists of smooth muscle (internal anal sphincter) which is innervated by the autonomic nervous system. In contrast, the outer tube (external anal sphincter) is made of striated muscle which is innervated by the somatic nervous system.

GROSS ANATOMY

The anal canal can be divided into three zones according to its epithelium. The proximal zone is covered by colorectal mucosa, which is identical with the rest of the colorectal mucosa. The distal zone, lined with squamous epithelium, is situated below the dentate line and extends outwards to the perianal skin. The intermediate zone, also called the anal transitional zone, extends in most people from the dentate line and about 1 cm proximally. Its extent may vary

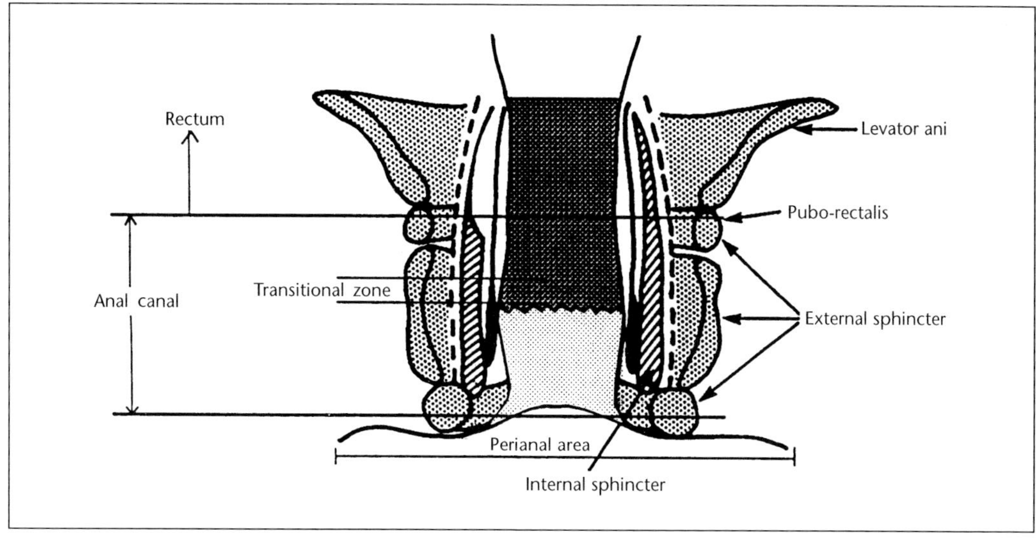

Fig. 1.1. *Schematic representation of the anal canal and it's relation to the anal margin and rectum.*

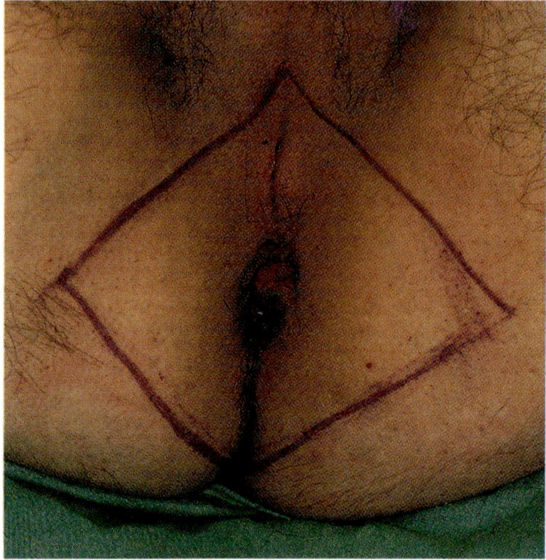

Fig. 1.2. *The Perianal skin area.*

from 0.6 to 2 cm proximal to the dentate line. The dentate line is situated about 2 cm proximal to the anal verge (Fig. 1.1), about halfway up the anal canal, and this marks the site of the anal membrane in the fetus. As a result of the rectum's narrowing towards the anal canal, 12–14 longitudinal folds are created (Fig. 1.3).

These folds are called the anal columns of Morgagni. Small crypts are located between these columns at their distal ends. These crypts are called the anal valves and the anal glands (anal sinuses) empty into them. This region is of interest since blockage of the anal sinuses can result in infection with abscess formation and sepsis.

The microscopic appearance of the colorectal zone of the anal canal is of the same red color as the rest of the rectum. Immediately proximal to the dentate line in the transitional zone, the color is dark-red to purple because of the underlying superficial veins. The distal zone (zone of squamous epithelium) looks like skin despite being thinner, smoother, and paler because it lacks accessory skin structures such as hair, fat, and sweat glands. This area of skin immediately below the dentate line extends about 2 cm distally and is called anoderm or the squamous epithelial zone, before it continues into the normal, thick, hairy perianal skin. The skin around the anal opening is especially rich in apocrine glands which can become sites of infection (hidradenitis suppurativa). The mucosa proximal to the dentate line is loose, while that

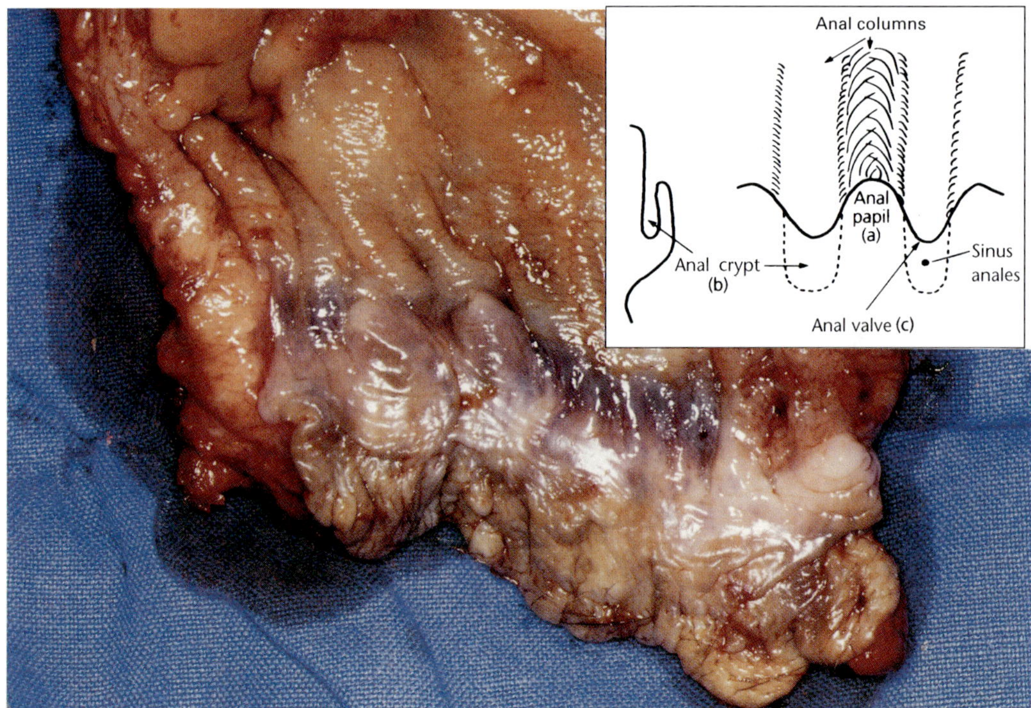

Fig. 1.3. *The anal canal (inset: a schematic representation of anal columns, valves and sinuses.*

overlying the dentate line and distal to it is tightly bound to the underlying tissue.

A layer of fibromuscular tissue separates the anal canal from the coccyx posteriorly and anteriorly from the membranous part of the urethra and bulb of the penis in men and from the lower end of the vagina in women. On each side of the anal canal are the ischiorectal fossa.

HISTOLOGY

The proximal zone of the anal canal is covered by the same columnar epithelium as seen in the colorectal mucosa and is, therefore, called the colorectal zone. The histological picture of the transitional zone, which represents the correlations of endoderm or ectoderm in fetal life, is in contrast variable. Depending on age a gradual change from columnar epithelium to transitional epithelium (as in the urinary tract) to squamous epithelium is found here. There are normally four to ten intermuscular glands in the anal canal (Fig. 1.4), which, via the anal ducts, empty into the anal crypts. The glands have multilayer squamous epithelium near their openings, deeper lined by transitional epithelium or columnar epithelium. Up to two glands may empty into each crypt, but not every crypt has an anal gland emptying into it. The glands extend through the internal anal sphincter into the intersphincteric layer. They never, however, extend through the external anal sphincter. The glands are thought to play a

role in the formation of anal abscesses and fistulae. They are also an extremely rare origin of extramucosal adenocarcinomas.

The anoderm or squamous epithelial zone consists histologically of non-hornified stratified squamous epithelium and the perianal skin consists of normal hornified stratified squamous epithelium.

ANAL CANAL MUSCULATURE

The sphincter musculature is composed of circular and vertical fibers and regulates the defecation mechanisms. The circular muscle is made of an internal layer – the internal anal sphincter – and an external layer – the external anal sphincter. Similar to the circular muscle, the vertical muscle consists of an internal and an external layer. The internal layer lies between the internal anal sphincter and the mucosa of the anal canal and is called the muscularis submucosae ani. Between the internal and external anal sphincters lies the external region of the two vertical muscle layers which is called the longitudinal intersphincteric muscle layer. Fig. 1.5 shows a schematic representation of the sphincter muscles.

The internal anal sphincter

The internal anal sphincter can easily be recognized by its white muscle fibers. The muscle has involuntary regulation. It is a flat, tubular muscle with circular and diagonal smooth muscle fibers that are gathered in discrete bands. The muscle is a continuation of the rectum's smooth musculature. It is about 2.5 to 4 cm long and about 0.5 cm thick.

The external anal sphincter

The external anal sphincter is more voluminous and consists of striated muscle. It is a voluntary muscle. It surrounds the length of the internal sphincter and lies completely superficial at the anal verge. The separation of these two muscles is often visible at this point and can be seen as Hilton's white line which corresponds to the transition between skin and mucosa. The term Hilton's white line is, however, not based on any actual anatomical structure and many do not use the name any longer. The cleavage

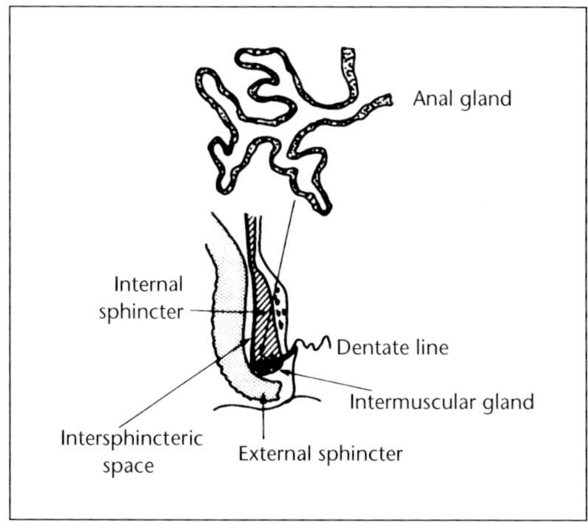

Fig. 1.4. *Anal glands and their outlet in the anal crypts.*

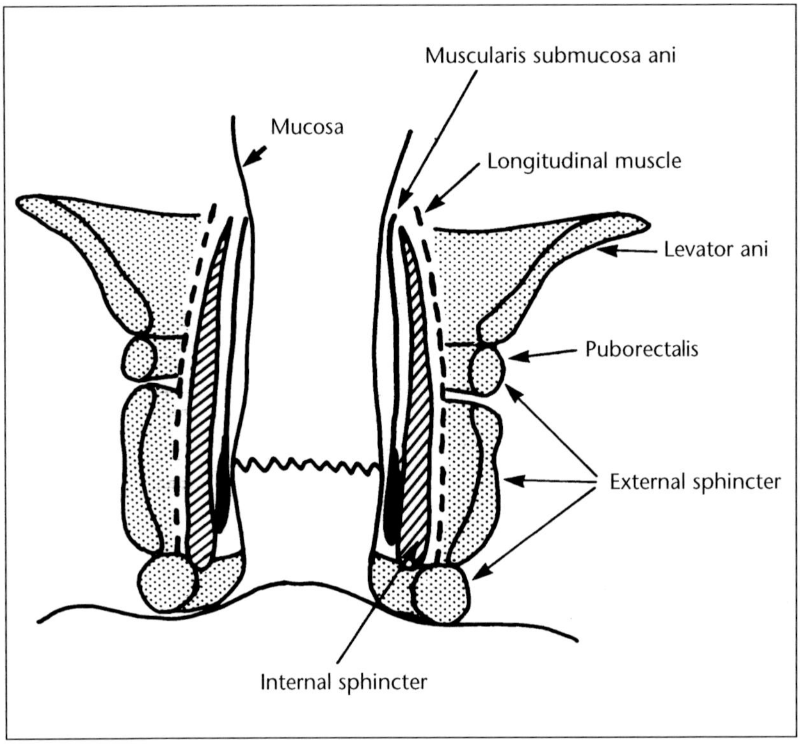

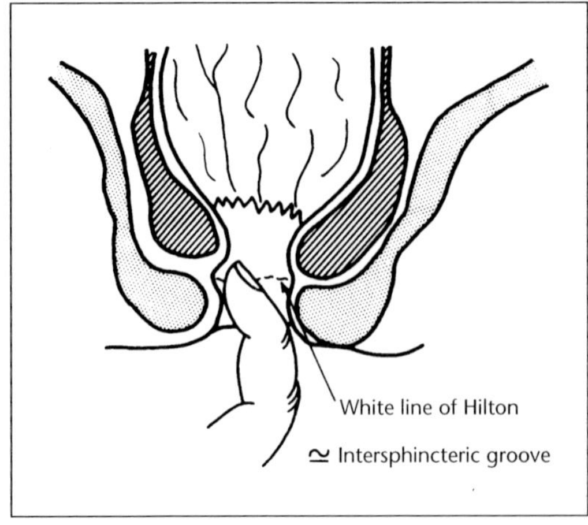

Fig. 1.5. *The muscles of the anal region.*

between the internal and external sphincter can often be palpated as an intersphincteric groove (Fig. 1.6), which is of significance when undertaking a lateral subcutaneous sphincterotomy. The muscle is often divided into three regions: subcutaneous, superficial, and deep (Fig. 1.7). The deep fibers have no actual insertion but continue into the puborectalis muscle. In contrast to this the superficial fibers are attached posteriorly to the anococcygeal ligament and anteriorly to the central tendon of the perineum. The subcutaneous fibers insert into the subcutaneous layer of the skin and can be easily palpated externally. Proximally the external muscle fibers are in continuity with the levator ani muscles. Posteriorly the muscle is attached to the coccyx and anteriorly to the perineal body.

The external muscle is sometimes described as a triple-loop system with an uppermost loop consisting of the puborectalis muscles and the deep fibers of the external sphincter, a middle loop made of the superficial parts of the external sphincter, and finally a distal loop arising from the subcutaneous parts of the external sphincter (Fig. 1.8). Coordination of these loops ensure the airtight closure of the anal canal.

Corrugator cutis ani muscle

This is a small tubular muscle which inserts proximally to the intersphincteric line and fans out distally to the skin. The radial folds of the skin around the anus are caused by this muscle

Fig. 1.6. *The white line of Hilton and the intersphincteric groove.*

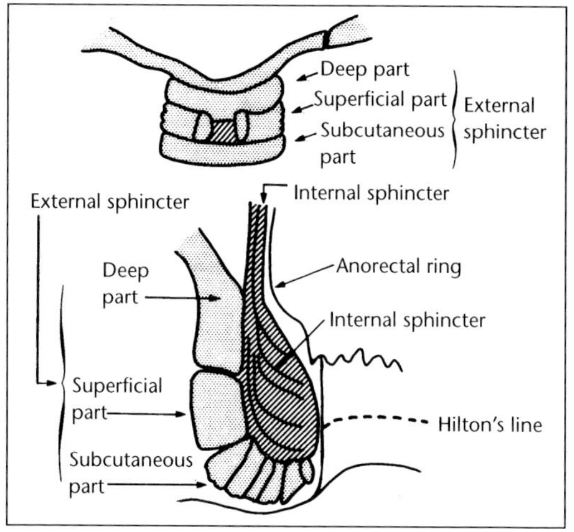

Fig. 1.7. *External sphincter and its subcutaneous, superficial and deep parts.*

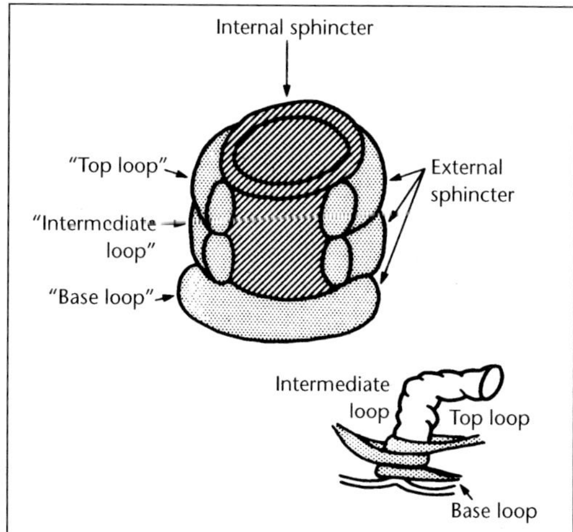

Fig. 1.8. *M. Sphincter ani externus and its "Triple loop" system that secures airproof closure of the anal canal.*

which can be regarded as a remnant of the panniculus carnosus.

Levator ani muscle

The levator ani muscle, which is also known as the anorectal ring, constitutes the pelvic diaphragm together with the coccygeus muscle (Fig. 1.9). The muscle consists of two parts, the pubococcygeus muscle and the ileococcygeus muscle. The pubococcygeus muscle can be divided into four sets of fibers, namely levator prostatae, pubovaginalis, puborectalis,

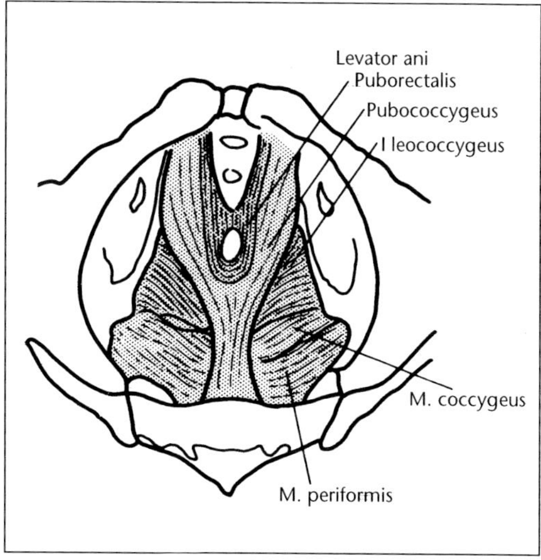

Fig. 1.9. *Levator ani and the coccygenous muscles close the base of the pelvis. They consist of several muscles.*

and the actual pubococcygeus. The puborectalis arises from the posterior part of the pubis runs posteriorly and together with the contralateral muscle forms a U-shaped loop, which is located at the transition between the rectum and the anal canal. The other muscle in the floor of the pelvis, ileococcygeus, also sends fibers to the rectum. By contraction the levator muscles contract and lift the pelvic floor. During defecation the muscles relax and the pelvic floor drops. By contraction of the sling made up of puborectalis, the lumen of the bowel is converted into a narrow cleft which prevents the passage of feces. The muscle is, therefore, of great significance in the maintenance of normal continence. The puborectalis muscle is responsible for creating an angle between the anal canal and the rectum (Fig. 1.10), which measures 80° in normal persons.

Muscularis submucosae ani

As mentioned, longitudinal smooth muscle fibres are located between the internal sphincter and the mucosa of the anal canal. They represent a continuation of fibers from the rectum. At the dentate line these fibers together with the internal sphincter create a ligament (Fig. 1.5) which binds the mucosa tightly to the internal anal sphincter. This is called Park's ligament.

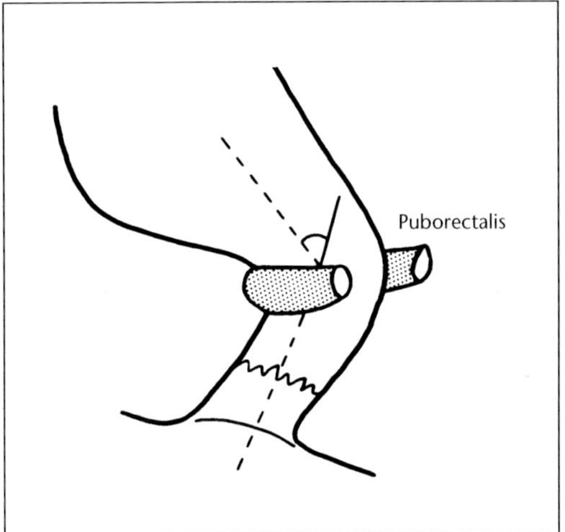

Fig. 1.10. *The puborectal muscle and the anorectal angle of 80° – normal.*

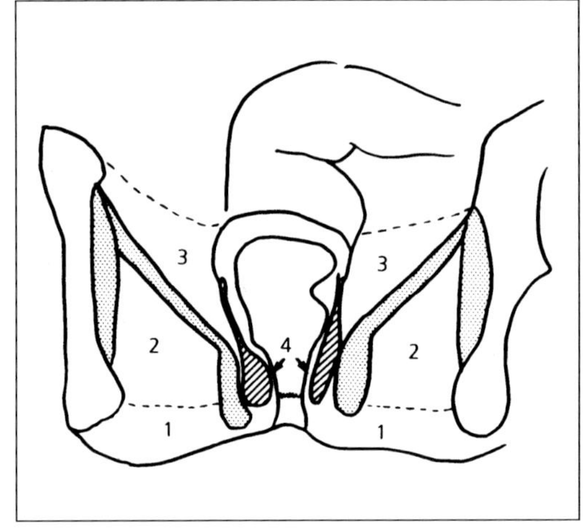

Fig. 1.11. *Anatomical regions in and around the anal canal. 1. Perianal space; 2. Ischiorectal space; 3. Supralevator space; 4. Submucous space.*

The longitudinal intersphincteric muscle layer

This muscle has a complex structure formed by continuation of the fibers from the rectum's longitudinal layer, striated fibers from the levator ani muscles, and aponeurotic fibers from the pelvic fascia.

TOPOGRAPHY OF THE ANAL CANAL

There are a number of well-defined anatomical regions in and around the anal canal that have a surgical relevance (see Chapter 7). Knowledge of these regions and especially their localizations and borders is, therefore, necessary.

The ischiorectal fossa

The ischiorectal fossa is a wedge-shaped region that lies below the pelvic diaphragm on either side between the anal canal and the wall of the pelvis (Fig. 1.11). Medially the fossa is limited by the external anal sphincter and higher up by the levator ani muscles. The lateral border is that part of the internal obturator muscle that lies beneath the tendoneous arch of the levator ani. Anteriorly the two fossae are separated from each other by the rectum and prostate, while posteriorly the only separation is the anococcygeal ligament. These regions

have a clinical interest as that they can be the focus of inflammatory processes arising from the anal canal.

The perianal space

This region is located just distally to the anal canal (Fig. 1.11) and continues laterally into the subcutaneous fatty tissue of the buttocks. Medially it extends along the lower part of the anal canal to the dentate line and continues into the intersphincteric space. This region contains the lowest part of the external anal sphincter, the inferior hemorrhoidal plexus, and branches from the inferior hemorrhoidal artery as well as lymph channels.

The intersphincteric region

This region is a potential space between the two sphincter muscles and continues downwards to the perianal space (Fig. 1.11).

The supralevator region

This region lies on either side of the rectum (Fig. 1.11) and is bounded distally by the levator ani muscles, medially by the rectum, laterally the wall of the pelvis, and proximally the peritoneum. Abscesses in this region arise from infection in the anal glands or spread from processes in the pelvis.

The submucosal region

This region is located between the internal anal sphincter and the mucosa (Fig. 1.11). Distally the region is limited by the dentate line and proximally it disappears into the submucosa of the rectum. The region contains the internal hemorrhoidal venous plexuses. Abscesses occasionally occur in this region.

The postanal region

There are two postanal regions: a superficial and a deep. The superficial region lies under the anococcygeal ligament (b) and connects the two ischiorectal fossae with each other behind the anus. The deep region or Courtney's retrosphincteric region (a) lies beneath the levator ani muscle (Fig. 1.12). This region also connects the two ischiorectal fossae with each other and usually horseshoe abscesses occur through this space.

BLOOD SUPPLY AND INNERVATION OF THE ANAL CANAL

Arterial supply

The blood supply of the anal canal appears to be more complex and variable than previously thought. It comes from three large arteries and one small.

The superior hemorrhoidal artery

This artery is the continuation of the inferior mesenteric artery which lies in the sigmoid

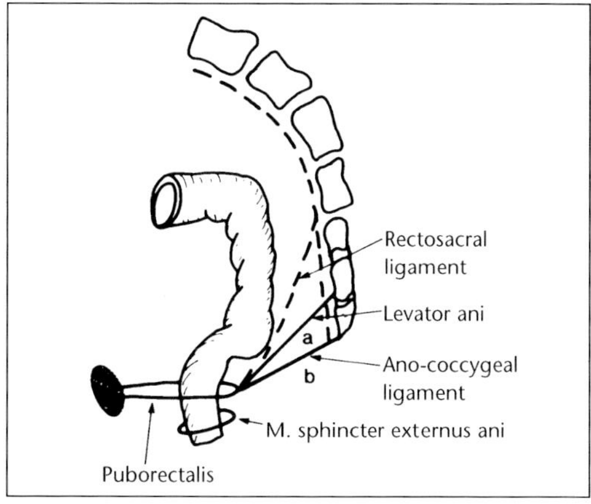

Fig. 1.12. *The postanal region.*

mesocolon (Fig. 1.13). It was previously thought that this artery divided into left and right branches in front of the third sacral vertebra, and that the right branch divided again into anterior and posterior branches. This pattern cannot, however, be confirmed in all studies. The artery runs into the rectum about 7 cm proximal to the anal canal. Its terminal branches supply the anal canal's submucosa, and end in the anal valves as a capillary plexus.

The middle hemorrhoidal artery

This artery rises from the anterior branch of the internal iliac artery (Fig. 1.13), and courses into the rectum anterolaterally at the level of the levator ani muscles. The artery supplies the proximal part of the anal canal.

Inferior hemorrhoidal artery

The main blood supply of the anal canal comes from the inferior hemorrhoidal artery which arises from the internal pudendal artery in Alcock's canal (Fig. 1.13). The artery runs through the ischiorectal fossa and branches, especially towards the sphincter muscles.

The middle sacral artery

This artery is of little significance in the blood supply of the anal canal (Fig. 1.13). It is, however, of great surgical importance since it can cause severe bleeding if it is torn during excision of the rectum. The artery runs in front of the sacrum, behind the presacral nerves and the superior hemorrhoidal artery to the distal part of the rectum and the proximal part of the anal canal.

It is controversial to what extent these four arteries anastomose with each other.

Venous drainage

Corresponding to the four arteries there are veins which drain blood partly to the portal system and partly to the systemic system (Fig. 1.14).

The superior hemorrhoidal vein drains the proximal part of the anal canal to the portal system via the inferior mesenteric vein. The middle hemorrhoidal vein drains the proximal part of the anal canal to the systemic circulation via the internal iliac vein. Finally the lower part

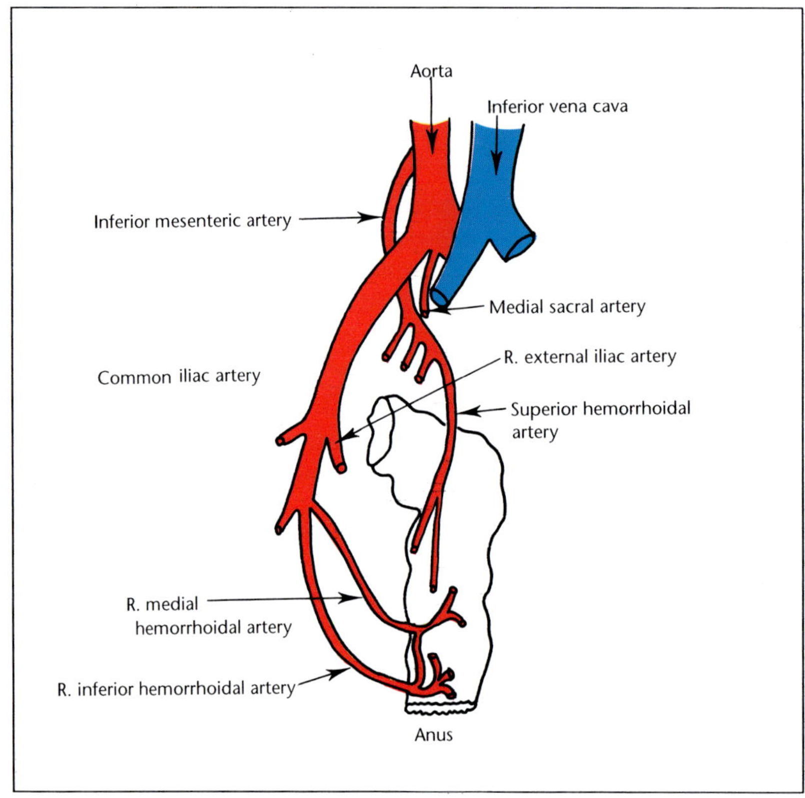

Fig. 1.13. *The arteries to the anal canal.*

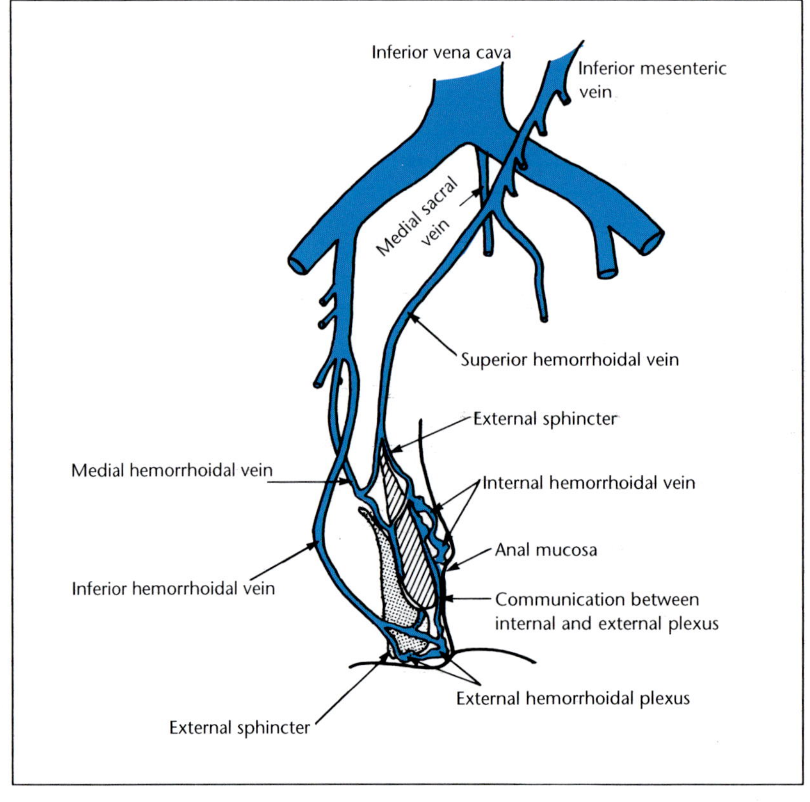

Fig. 1.14. *Veins from the anal canal.*

of the anal canal is drained by the inferior hemorrhoidal vein into the internal pudendal vein and then into the systemic circulation via the internal iliac vein. It is also controversial to what extent the veins anastomose with each other.

Lymphatic drainage

The lymphatic drainage from the anal canal which lies above the dentate line is complex. There is an intramural lymphatic system that connects the lymphatic channels in the anal canal with those in the rectum. The lymphatic vessels drain proximally to the inferior mesenteric lymph nodes and laterally via the middle hemorrhoidal lymphatic channels to the internal iliac lymph nodes (Fig. 1.15). Lymph from the part of the anal canal that lies distally to the dentate line usually drains into the inguinal lymph nodes, but drainage to the ischiorectal lymph nodes occurs in the presence of obstruction. This means that malignant growths situated in both the anal canal and the perianal region can metastasize to lymph nodes in the inguinal region and to intraabdominal lymph nodes along the large vessels.

Innervation

The anal canal as with most other organs is innervated by sympathetic and parasympathetic nerve fibers.

Sympathetic and parasympathetic innervation

In the lumbal region the sympathetic chain lies on the sides of the vertebral bodies (Fig. 1.16) in front of the psoas muscles. In the pelvis they lie on the pelvic fascia of the sacrum. Distally the sympathetic chains converge from the two sides to meet as an unpaired ganglion. Threads are given off to the abdominal aorta and its various branches in the pelvis. The sacral part of the sympathetic chain is connected to a large perivertebral plexus called the hypogastric plexus. This plexus forms a continuation of the celiac plexus and may be divided into an upper unpaired part and a lower paired part which is called the inferior hypogastric plexus. This lies

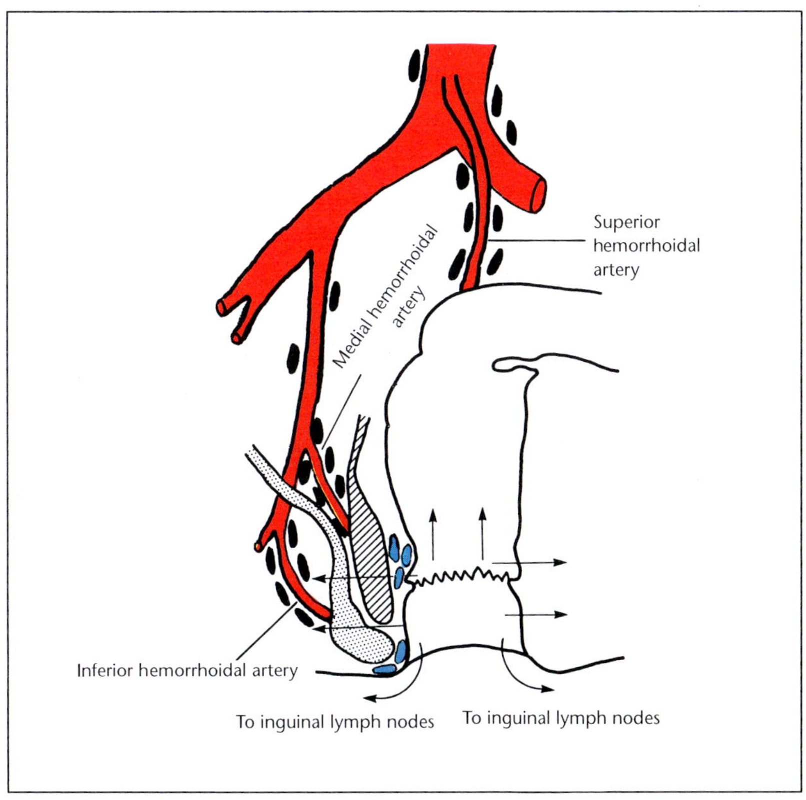

Fig. 1.15. *The lymph drainage of the anal canal.*

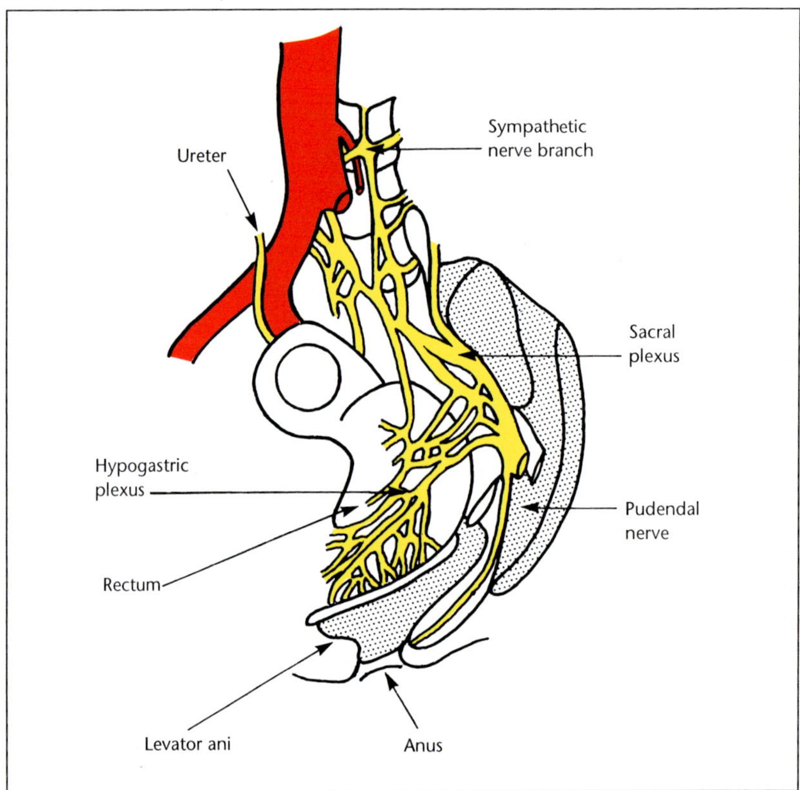

Fig. 1.16. *The lumbar sympathetic branch and innervation of the anal region.*

on each side of the pelvic cavity in the connective tissue posterolateral to the rectum. The plexus is formed from sympathetic fibers from the aortic plexus and from the lumbal and sacral chains. Together with the parasympathetic fibers (nervi erigentes) they form a plexus in the pelvis. Sympathetic fibers that are loosely connected to the rectum's posterolateral attachments can be easily damaged during resection of the rectum unless special care is taken.

The parasympathetic fibers are made up of the nervi erigentes which arise from the second, third, and fourth sacral nerves (Fig. 1.17). They run in the side wall of the pelvis together with the sympathetic fibers, as mentioned above. It should be stressed that both the sympathetic and parasympathetic nerve fibers are involved in erection and that damage to these nerves during resection of the rectum can cause incomplete erection, absence of ejaculation or total impotence.

The transmitters of the sympathetic and parasympathetic nervous system are the classic ones, such as noradrenaline or acetylcholine. Recent investigations however have shown a non-adrenergic and non-cholinergic autonomic nervous system – the peptidergic nervous system – where the transmitters and/or modula-

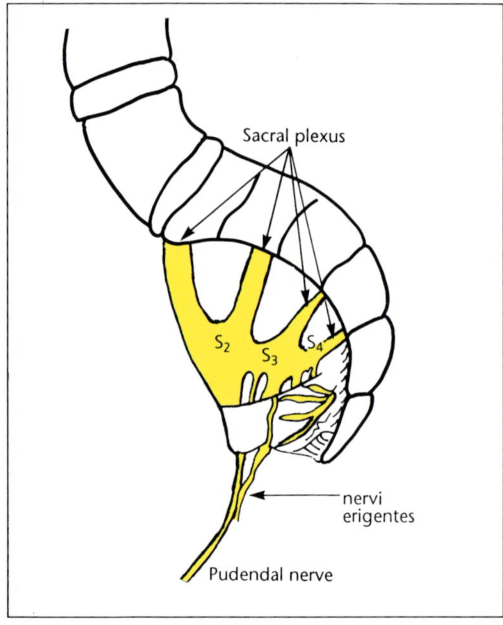

Fig. 1.17. *Nerve supply to the anus (lateral view).*

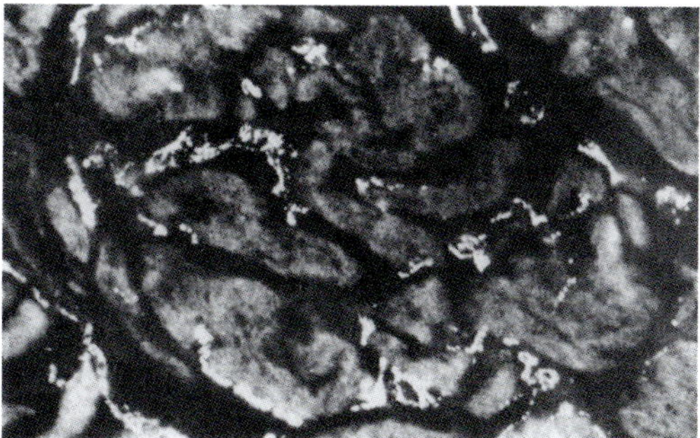

Fig. 1.18. *VIP nerves in the external anal sphincter.*

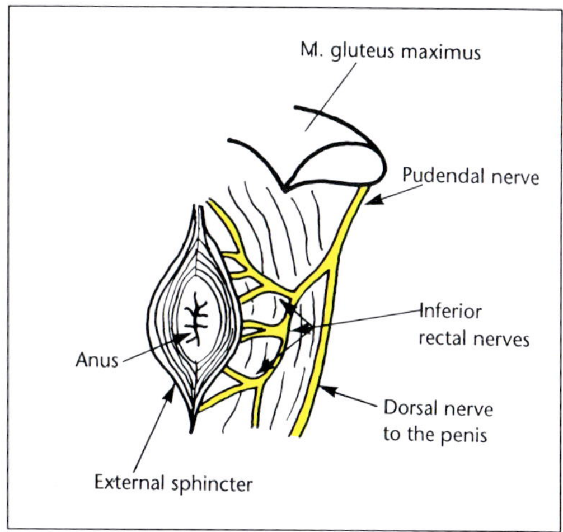

Fig. 1.19. *External sphincter.*

tors are the so-called regulatory polypeptides. Immunocytochemical investigations have amongst other things demonstrated nerves with the following regulatory peptides in the anal canal: vasoactive intestinal polypeptide (VIP), substance P (SP), enkephalin, neuropeptide Y (NPY), and galanine. Fig. 1.18 shows an immunohistochemical picture of VIP distribution in the different layers of the anal canal. The physiological significance of these peptides has not yet been ascertained. SP is thought to stimulate the internal anal sphincter while VIP relaxes it.

The anal canal. The sympathetic fibers stimulate and the parasympathetic fibers inhibit the internal anal sphincter. The external anal sphincter is innervated by the inferior rectal nerve which arises from the internal pudendal nerve (Fig. 1.19), and via a branch of the fourth sacral nerve. The levator ani muscle is similarly supplied from the fourth sacral nerve together with branches of the inferior rectal nerve.

The perianal area. The sensory fibers to the perianal skin and anoderm also arise from the inferior rectal nerve. The area above the dentate line, the transitional area, is not completely without sensation. The unpleasant feeling caused by, for example, application of a rubber band around an internal hemorrhoid probably arises from intraepithelial nerves in the transitional zone.

PHYSIOLOGY

The physiology of the anal canal is extremely complex and there are still many unresolved aspects even though the improved manometric electrophysiological and radiological methods of recent years have significantly improved our understanding of the normal function of the anus. Developments in peptide search have brought about a new and enhanced understanding of the function of smooth muscle and will without doubt contribute to a better definition of the physiology behind the two important functions of the anal canal, namely fecal continence and defecation.

Fecal continence

The precise mechanisms which contribute to the maintenance of normal fecal continence are not yet fully elucidated, so it is not possible to give an adequate definition of the concept. It is, however, clear that continence is maintained by a combination of various factors and that it is dependent amongst other things upon rectal capacity or compliance, integrity of the sphincter apparatus, sensory aspects, various reflexes, the anorectal angle, which is created by the puborectal muscle (see Fig. 1.10), and the consistency of the stool.

Rectal capacity

The rectum supplies a mechanical and a physiological reservoir function. The rectum is normally collapsed and the lateral angulation of the sigmoid colon (Fig. 1.20) as well as presence of the so-called Houston's valves (Fig. 1.20) contribute to a delay in fecal progression. Moreover, the weight of the feces accentuates these angles in the bowel and thereby their barrier effect. The physiological reservoir function is produced by the fact that the motorial activity in the rectum is much more marked than in the sigmoid colon with more frequent and more powerful muscle contractions. This high pressure zone in the rectum functions as a barrier against movement of feces in a distal direction. Similar pressure differences between the upper and lower parts of the anal canal accentuate this pressure barrier (Fig. 1.21 and 1.22), and this activity in the anal canal has a great significance for the maintenance of continence, especially for flatus.

Integrity of the sphincter muscles

The sphincter muscles are of great significance for the maintenance of continence of feces and flatus. Manometric studies have shown there is a resting high pressure zone in the anal canal which is an effective barrier against pressure in the rectum. The resting pressure in the anal canal lies between 24 and 125 mm Hg while the resting pressure in the rectum is only 5 to 25 mm Hg. This pressure zone extends normally 3 to 7 cm from the anal verge with the highest pressure lying about 2 cm from the dentate line. Both the external

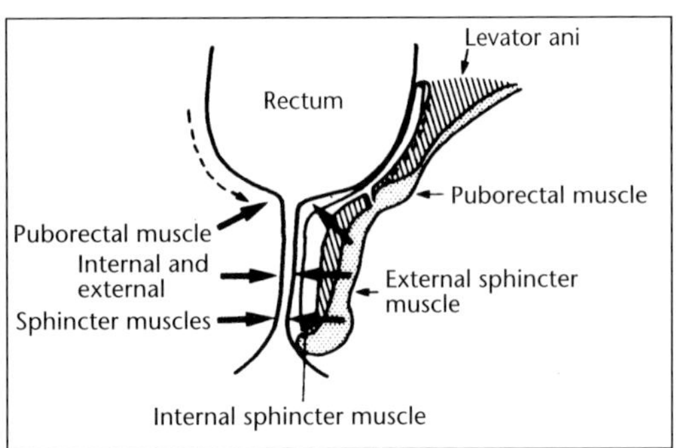

Fig. 1.21. *Frontal pressure barrier of the anal canal.*

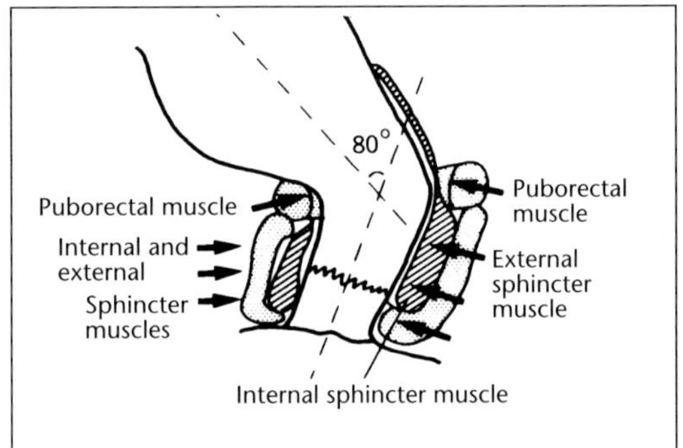

Fig. 1.22. *Lateral view of the pressure barrier of the anal canal.*

and the internal sphincters contribute to the maintenance of resting pressure (Fig. 1.21 and 1.22).

In contrast to other striated muscles, the external sphincter is not inactive under resting conditions and it does not degenerate after denervation. The tone in the levator ani and in the external anal sphincter is in constant flux even during sleep. Deep breathing and the Valsalva maneuver cause a severe activity increase in the muscles with a resulting pressure increase in the anal canal. The internal anal sphincter is thought to play a big role in the maintenance of the resting high pressure zone in the anal canal. It has not yet been established which of the two sphincters has the greatest significance in the maintenance of continence. The internal anal sphincter seems to play the greater part in the maintenance of resting pressure since paralysis of the external anal sphincter does not change the pressure significantly. The pressure in the anal canal is dependent not only on the tone of the sphincters but also on their ability to oppose opening of the anal canal.

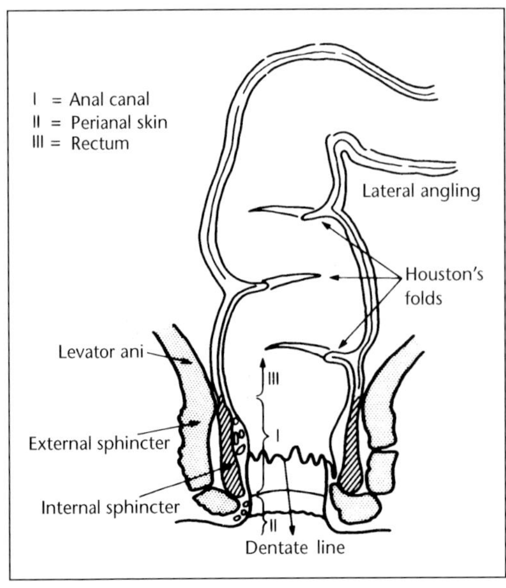

Fig. 1.20. *Lateral angling and Houston's folds.*

Sensory mechanisms

The significance of the sensory mechanisms in fecal continence have not been defined but much suggests that rectal sensation is an important signal for the feeling that feces are on their way to the rectum. Anal sensation contributes to the perception of what kind of material is involved. According to Duthie & Gairns, the sensory nerve endings in the anal canal are conventional nerve endings that register pain (lying freely intra-epithelially), touch (Meissner's bodies), cold (Krause's bodies), pressure (Paccini bodies), and friction (genital bodies). There are in addition unnamed nerve endings which are responsible for the differentiation of sensory stimuli. There are no receptors in the rectum. The rectum is only sensitive to stretch. It is thought that the intrinsic nerve endings do not play a role in continence but that they are merely warning systems for impending defecation. The puborectalis muscles and the levator ani are thought to be the most important factors for maintenance of fecal continence. Extrinsic sensory receptors have been found in these muscles. Research indicates that the puborectalis muscle has both motor and sensory functions that are essential for continence. The puborectalis is a much more sensitive receptor to pressure changes than the rectum itself. It is now clear that an intact rectum is not necessary for sphincter activity.

The puborectalis muscle

The most important mechanism for the maintenance of normal fecal continence is without doubt the angle between the anal canal and the rectum which exists because of the constant tonic activity in the puborectal sling (Fig. 1.10 and 1.21). The 80° angle between the axes of the anal canal and the rectum is present except during defecation and when the hips are flexed to more than 90°. Over and above the effect of the puborectalis muscle in the maintenance of continence there are other factors which support this effect. On the basis of manometric and radiological studies it has been suggested that an increased intraabdominal pressure at the level of the levator ani muscles at the anorectal ring can compress the anal canal (flutter valve) (Fig. 1.23). Another theory, the so-called flap valve theory, states that continence occurs because the anterior rectal wall creates a valve at

the level of the anorectal ring (Fig. 1.24). Defecation occurs when the valve disappears as a result of relaxation of the puborectal muscle, whereby the anorectal angle disappears. Video recordings of Valsalva's maneuver have, however, shown that the anterior rectal mucosa does not create such a valve and furthermore that a postanal repair, which is designed to recreate the angle between the anal canal and the rectum, is successful in about 60% of cases even though the angle is not reconstituted. The significance of the puborectal muscle for the maintenance of continence is thus still controversial. Other mechanisms of significance for maintenance of anal continence are different tension effects proximally in the anal canal. The tension effect is maximal posteriorly, less laterally, and minimal anteriorly. This suggests a combined effect of the internal anal sphincter

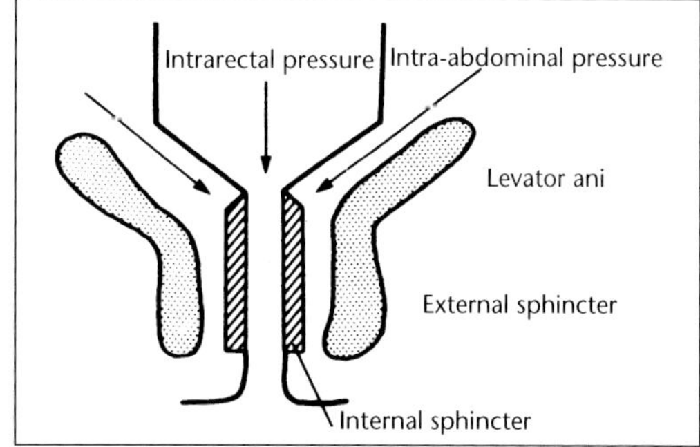

Fig. 1.23. *The flutter valve theory.*

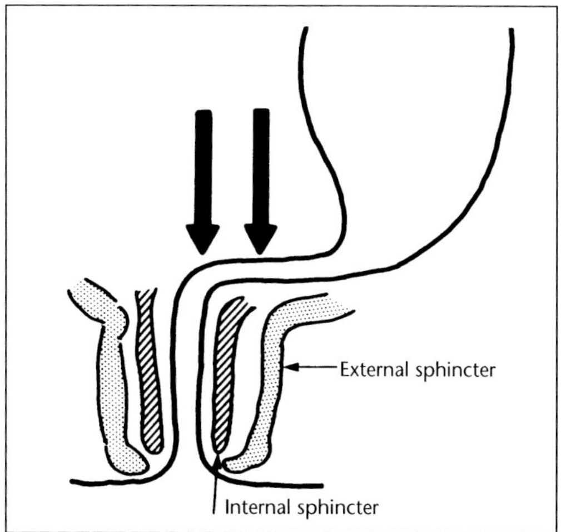

Fig. 1.24. *The flap valve theory.*

and puborectalis. This so-called triple loop system (Fig. 1.8) should ensure continence for flatus and that the upper loop pulls upwards and forwards, the middle loop horizontally and posteriorly while the lower loop pulls downwards and forwards. If one of these loops fail, the patient will still be continent for feces but not for flatus.

Defecation

The desire for defecation is perceived only when feces are passed into the rectum by peristaltic activity and have caused distension of the rectum. As long as the rectum is empty no defecation urge occurs. As mentioned earlier, the external and internal anal sphincters prevent feces, which have accumulated in the rectum, from slipping out. The defecation process is partly voluntary and partly involuntary. When distension of the rectum reaches a certain volume and constitutes a sufficient stimulus, the colon contracts and the anal sphincter relaxes resulting in evacuation of the feces. The reflexes concerned involve only the sacral part of the spinal cord. This part of the process is under voluntary control and involves relaxation of the external anal sphincter and probably simultaneous contraction of the diaphragm and abdominal muscles. These processes occur many times a day but the results depend upon the circumstances occurring at the time. They can be repressed by a complex cortical inhibition of the reflexes from the anal canal. Normally rectal distension causes a relaxation of the internal sphincter which in turn gives rise to a contraction of the external sphincter in an attempt at maintaining continence.

If one decides to defecate a sitting position with flexion of the hip joints is assumed whereby the angle between the anal canal and rectum disappears. One performs a Valsalva maneuver which overcomes the pressure of the external anal sphincter and the increased pressure in the rectum. At the same time, inhibition of the external anal sphincter allows feces to pass through the anal opening. Thereafter, the muscles relax and the anal canal closes again. If a rectal balloon is filled with about 10 ml of water, a transitory increase in the electromyographical activity of the external anal sphincter occurs, in contrast to the internal anal sphincter

which shows a corresponding reduction in activity. With persistent filling of the balloon the increased pressure in the rectum is maintained for a few minutes whereafter it falls to basal levels. This is called the adaptation response (Fig. 1.25). The afferent nerve endings for the adaptation reflex are found in the ampulla of the rectum and in the levator ani muscle. Another response is the so-called sampling response (Fig. 1.25), which consists of a transitory relaxation of the internal anal sphincter allowing contents of the rectum to come into contact with the somatic sensory lining of the anal canal whereby one can perceive the nature of the rectal content. By increasing the intraabdominal pressure by voluntary control while simultaneously maintaining the activity of the external anal sphincter, flatus is allowed to pass while solid content remains in the ampulla.

Manometry

In the past, manometric measurements have been essential for the physiological and pathophysiological evaluation of the function of the anal canal. Many different systems are used, which make comparisons difficult. The methods are relatively gross and reflect only the activity of the smooth muscle, i.e., the function of the internal anal sphincter. Examination of the striated muscles are undertaken by electromyography (EMG examinations). Conventional manometry is carried out as a profile examination, for example, by means of an 8F pressure

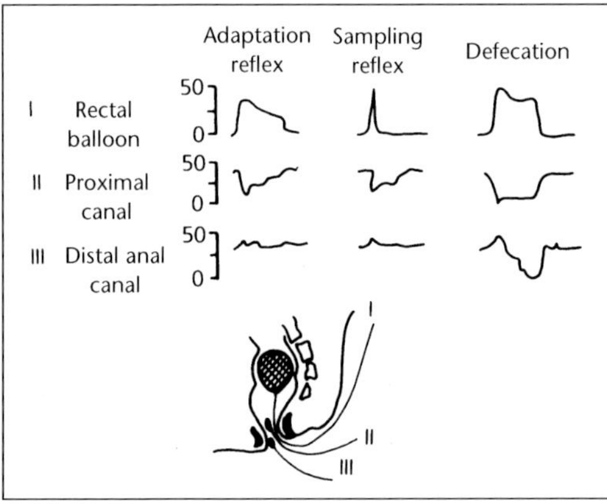

Fig. 1.25. *Adaptation and sampling responses.*

catheter with side holes. The pressure catheter is connected via a manometer to a pressure transducer and stylus. The pressure catheter is perfused with saline solution at 2–3 ml per second and is pulled through the anal canal at a constant speed (2–3 mm per second). A pressure profile can thus be drawn. The perfusion catheters must be of the low compliance type with a DP/DT not greater than 800 mm Hg per second. Moreover, it is important that one always states precisely which method has been used (catheter dimension, continuous or intermittent retraction of catheter, localization of the side holes). One measures the length of the high pressure zone of the anal canal, the maximal resting pressure as well as the maximal squeeze pressure. The pressures should be given in kilo-Pascals. Each laboratory has its own normal values. The high pressure zone of the anal canal varies from 2 cm to 2.5 cm depending on the sex of the subject. The examination is carried out during rest under maximal contraction. From the curve (Fig. 1.26), the maximal resting squeeze pressure as well as the length of the pressure zone of the anal canal can be found.

Profile studies can also be undertaken with a tip-transducer. It is important that the sensor of the transducer is placed in the same direction in every examination in order to obtain reproducible results. The tip-transducer is similarly pulled through the anal canal at a constant speed. Manometry is carried out with the patient in the supine or lateral position with the hips and knees flexed. To prevent irritation of the sphincter muscles the bowel is not cleansed before the investigation. One can also evaluate pressure relationships by digital palpation (the "educated finger"). This simple method is valuable in the evaluation of anal function but does not reflect the exact anal pressure. It is characteristic that anal manometry has a wide individual variation of up to 25%. It is important to take note of this if one wishes for example to assess the effect of a treatment using manometry. What one might regard as an effect may be due to the large normal variation in pressure relationships emerging from conventional manometry. There is a poor correlation between digital palpation and manometry. Anal manometry should be combined with functional radiological investigations and electromyography to benefit from the investigation. Taken alone it has a limited value except in the diagnosis of Hirschsprung's disease.

Sensory investigations

The significance of sensory reflexes in anal continence has not yet been elucidated properly. Investigations such as electrical sensitivity and temperature sensitivity are gross and have no practical clinical significance. The interpretation of the results of the investigations is not yet clear.

Electromyography

EMG is a recording of the striated muscles' electrobiological activity (Fig. 1.27). EMG is carried out by placing needle electrodes in the musculature or surface electrodes in the skin. The purpose of the investigation is to ascertain whether the electrical impulses in the external anal sphincter or in the puborectalis muscles are functioning correctly. One can thus register curves in patients with disorders of the sphincter apparatus and in this manner demonstrate whether there is actual paralysis and to a certain extent gain information concerning the degree of paralysis. The recordings can also tell whether the paralysis is muscular, i.e., arising in the muscle itself, or whether it is neurogenic.

The investigation is not used very often in the daily clinical evaluation of anal disorders but should be used more often in evaluation of anal incontinence or constipation (spastic perineal syndrome) together with radiological investigations. A useful method of evaluating the function of the somatic muscles is the

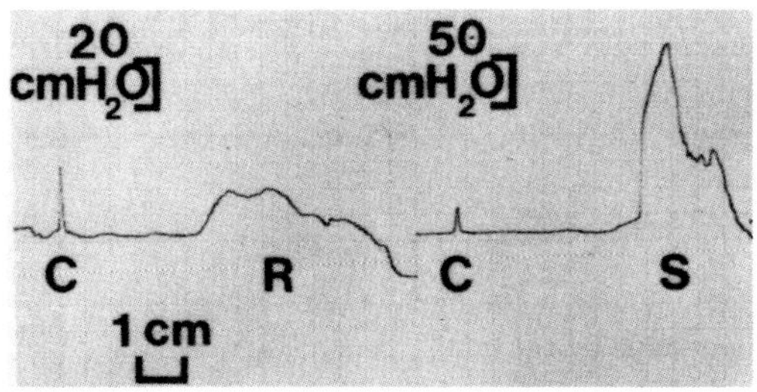

Fig. 1.26. *Manometric pressure responses in the anal canal. R = resting pressure. S = squeeze pressure.*

so-called single fiber EMG. Among other things, the investigation has shown that many patients with fecal incontinence are in fact suffering from a denervation injury of the striated muscle.

The anal reflex

The classic anal reflex is evoked by a stimulus supplied to the perianal skin or to the anoderm of the anal canal and consists of fasciculation of the perianal skin caused by contractions of the external anal sphincter. The reflex is mediated by spinal reflex connection between the perianal skin and the external sphincter. Reflex contraction of the external anal sphincter can also be evoked by stimulation of the glans penis, clitoris, rectal mucosa, urethra or bladder.

Technique

The anal reflex can be stimulated by many means. The reflex is normally evoked by stimulating the perianal skin with superficial electrodes and the effect analyzed by EMG investigation of the external anal sphincter. The reflex consists of an early short response and a late response of longer duration (Fig. 1.28). The anal reflex is always present in a normal subject and occurs about 15 ms after perianal stimulation. The primary short response of the anal reflex need not always be present and its absence is of doubtful significance.

The abnormal anal reflex

Defective reflex paths (nerve tracks) result in an absent anal reflex, increased latent period or reduction in the power of the reflex.

The lack of a late response in the anal reflex is seen in lesions of the medulla and general defects, for example, myelomeningocele or infections, tumors or trauma to the reflex paths.

Rectal compliance

This is investigated with an inelastic balloon filled with water or air placed in the rectum. The diameter of the balloon must be greater than that of the rectum. The balloon, containing an internal pressure catheter, is placed 10 cm from the anal verge. The balloon is distended during simultaneous measurement of the intrarectal pressure, $C = \Delta V/\Delta P$. High compliance suggests good reservoir function. The volume and the corresponding pressure are measured at four different situations:

1. first sensation of content in the rectum;

2. first sensation of urge for defecation;

3. impending urge for defecation;

4. the maximal tolerated volume that causes pain and discomfort.

The normal values are about 100 ml (50-300), 150 ml (100-350) and 225 ml (150-400) for first sensation of defecation urge, constant sensation of defecation urge and maximal tolerated volume, respectively.

The rectosphincteric reflex can be investigated with a balloon in the rectum which is inflated during simultaneous pressure measurement in the anal canal. With distension of the rectum, pressure in the anal canal normally falls. If the reflex is present one can exclude Hirschprung's disease. The reflex may be absent after a low anterior resection of the rectum and in patients with megarectum.

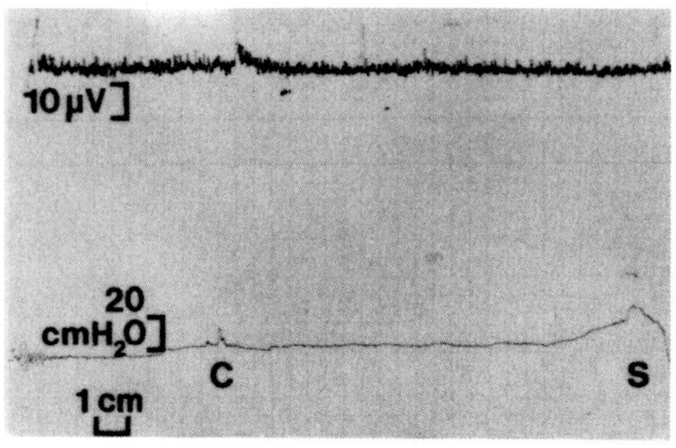

Fig. 1.27. *Electromyographic (EMG) registration of the activity of the external sphincter (upper curve).*

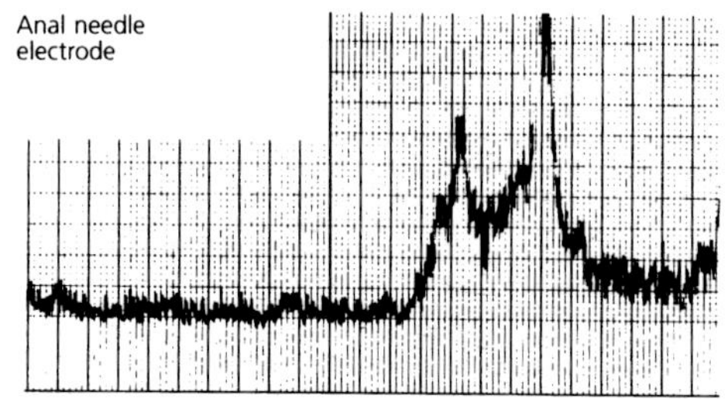

Fig. 1.28. *Registration of the anal reflex.*

The test is used in the investigation of constipation and fecal incontinence.

Distention test

Manometry gives information about pressure relationships in the anal canal but does not measure its elasticity. Elasticity is an important parameter that reflects tissue's ability to withstand distension. Cross-sectional diameter measurements in the anal canal can probably increase our understanding of continence and defecation mechanisms. It must, however, be accepted that both the collagen content and the thickness and strength of the muscle layer are important factors in the maintenance of normal continence.

The cross-sectional area is measured by means of a catheter attached to a balloon. The balloon can be distended by introducing a given pressure through an infusion canal attached to a fluid column using saline. The pressure within the balloon is measured by a perfused pressure system or with a strain gauge transducer. The cross-sectional area can be estimated from the measurements of the electrical impedants of the saline within the balloon. From two stimulation electrodes a constant alternating-current is generated (34 mA, 4 Hz). One or more pairs of detecting electrodes that lie between the stimulation electrodes measure the impedance of the intervening fluid. By using Ohm's law and the field gradient principle one can see that a linear relationship exists between the cross-sectional area and the measured impedance. From these measurements the anal canal's stiffness or rigidity (changes in pressure divided by changes in cross-sectional area) and compliance can be calculated.

The method is not uncomfortable if the pressure applied is less than 70 cm water. Investigations on healthy women have shown that rigidity is greatest distally in the anal canal and that it falls markedly in the more proximal areas of the anal canal. At the same time it has been observed that rigidity decreases during distension (up to 3 cm in diameter), i.e., an active relaxation occurs which is probably reflex mediated (Fig. 1.29).

This method has been used in the esophagus in animals. Human studies have been undertaken in the distal urinary tract and duodenum. We do not yet know the value of this method of investigation in the anal canal, but it seems to have interesting possibilities for the future.

SUPPLEMENTARY READING

Courtney H. Anatomy of the pelvic diaphragm and anorectal musculature as related to sphincter preservation in ano-rectal surgery. Am J Surg 1950; 79: 155.

Drummond H. The arterial supply of the rectum and pelvic colon. Br J Surg 1913; 1: 677.

Duthie HL. Defaecation and the anal sphincters. Clin Gastroenterol 1982; 11: 621.

Duthie HL, Gairns FW. Sensory nerve endings and sensation in the anal region of man. Br J Surg 1960; 47: 585.

Fenger C. The anal epithelium: a review. Scand J Gastroenterol 1974; 14: 114.

Fenger C. The anal transitional zone. Acta Pathol Microbiol Scand Sect A 1979; 87: 379.

Goyal RK, Rattan S, Jones W. VIP as a possible neurotransmitter of non-cholinergic non-adrenergic inhibitory neurones. Nature 1980; 288: 378.

Oh C, Kark AE. Anatomy of the external anal sphincter. Br J Surg 1972; 59: 717.

Pedersen E, Klemar B, Schrøder HD, Tørring J. Anal sphincter responses after perianal stimulation. J Neurol Neurosurg Psychiatry 1982; 45: 770.

Phillips SF, Porter NH. Some aspects of anal continence and defecation. GUT 1965; 6: 396.

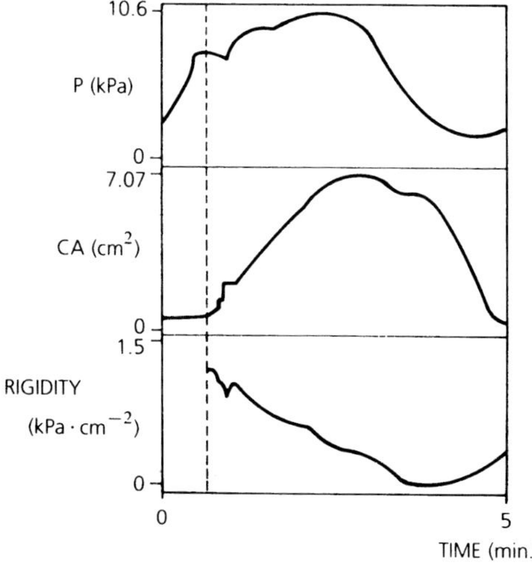

Fig. 1.29. *Example of simultaneous measuring of the intra-anal pressure (P) and cross sectional area (CA) 0.5 cm from the distal opening during balloon distention.*

Schuster MM, Hendrix TR, Mendeloff AI. The internal anal sphincter response; manometric studies on its physiology, neural pathways, and alterations in bowel disorders. J Clin Invest 1963; 442: 196.

Shafik A. A new concept of the anatomy of the anal sphincter mechanism and the physiology of defaecation. Invest Urol 1975; 12: 412.

Taverner D, Smiddy FG. An electromyographic study of the normal function of the external anal sphincter and pelvic diaphragm. Dis Colon Rect 1959; 2: 153.

Ustach TJ, Tobon F, Hembrecht T, Schuster MM. Electrophysiological aspects of human sphincter function. J Clin Invest 1979; 49: 41.

Wankling WJ, Brown BH, Collins CD, Duthie HL. Basal electrical activity in the anal canal in man. GUT 1968; 9: 457.

II. Symptomatology

The various symptoms of anal disorders occur frequently among the general population. In many cases these symptoms are caused by minor disorders which do not require treatment. This fact, however, often weakens the attention of both the patient and the doctor. Patients are slow in seeking a doctor's advice (patient delay) and the primary diagnosis is often wrong in patients presenting with a malignant disease (doctor's delay). This will of course have serious consequences for the patient. It should be stressed, therefore, that a malignant disease may be masked by apparently benign symptoms. The principle should be that any symptom of the anorectum should be regarded as arising from a malignant disease until a thorough examination of the area has disproved this. It is useless to write a prescription for an ointment to a patient with fresh bleeding from the anus in the belief that the bleeding is caused by hemorrhoids without first having undertaken a thorough anorectal examination.

BLEEDING

Bleeding is an important symptom in patients with anorectal diseases. One can come far with the differential diagnostic considerations merely from an exact history of the bleeding symptom. Symptoms of bleeding can be divided in the following manner:

A spray of bleeding on the toilet bowel

Fresh, red blood that drips into the toilet without mixing with the feces:

Hemorrhoids

Blood on the toilet paper

This indicates that the bleeding arises from a lesion of the anal verge. This is usually due to one of the following:

Prolapsed hemorrhoids
Anal fissure
Ruptured perianal hematoma
Ulcer of the anal verge or in the perianal skin (secondary syphilis, tuberculosis, and external fistula opening)
Tumor of the anal verge or of the perianal skin (benign, premalignant or malignant)

Bleeding after termination of defecation

In most cases this is due to hemorrhoids, but can also be seen in the following conditions:

Hemorrhagic proctitis
Radiation proctitis
Induced by medications (suppositories)

Streaks of blood on the feces

This is a sign of a localized lesion of a certain volume. The commonest cause is a polyp in the rectum. An anal fissure can also give this kind of bleeding when the sphincter is atonic rather than spastic.

Isolated rectal bleeding (Evacuation of clots)

Bleeding occurs independently of defecation. In most cases it is due to a lesion in the colon but may also be seen in the following situations: A lesion induced by a foreign body (e.g., thermometer) Spontaneous rupture of a rectal varix (rare).

Rectal bleeding with defecation disturbances

This is seen especially with villous adenomas in the rectum or with the bloody diarrhea associated with ulcerative colitis or Crohn's disease. Can also occur with a dysentery-like syndrome which may be caused by a malignant lesion of the sigmoid colon (commonest cause), diverticulosis of the colon (extremely rare) or with true dysentery.

Mucus mixed with blood

Most often this indicates a low-lying rectal carcinoma or one in the anal canal. Figure 2.1 shows the most frequent causes of rectal bleeding. It is important to ascertain whether one is dealing with red or black rectal bleeding since many conditions of the gastrointestinal tract manifest with rectal bleeding. Bleeding arising from lesions high in the gastrointestinal tract is most often present with black rectal bleeding (melena) unless the patient is bleeding massively. In such situations the patients will demonstrate signs of hypovolemia or be manifestly shocked.

PAIN

Most disorders of the anal region perhaps with the exception of non-prolapsing internal hemorrhoids are accompanied by some form of discomfort or pain of varying intensity. A precise history of the pattern of pain will often lead to a diagnosis, which can be confirmed by objective examination. Disorders of the perianal region can very often be distinguished from lesions of the anal canal in this manner. All patients presenting with perineal pain should undergo a rectoscopic examination since the pain can be an expression of a rectal cancer with nerve involvement.

Pattern of pain for perianal disorders:

Diffuse burning or stinging
Exanthema
Inflammation
Pruritus ani

Severe sharp pain in the anal verge
Abscess
Anal fissure
Perianal hematoma
Prolapsed incarcerated internal hemorrhoids

Dull aching pain (with swelling of the anal verge)
Abscess
Tumor
Perianal hematoma

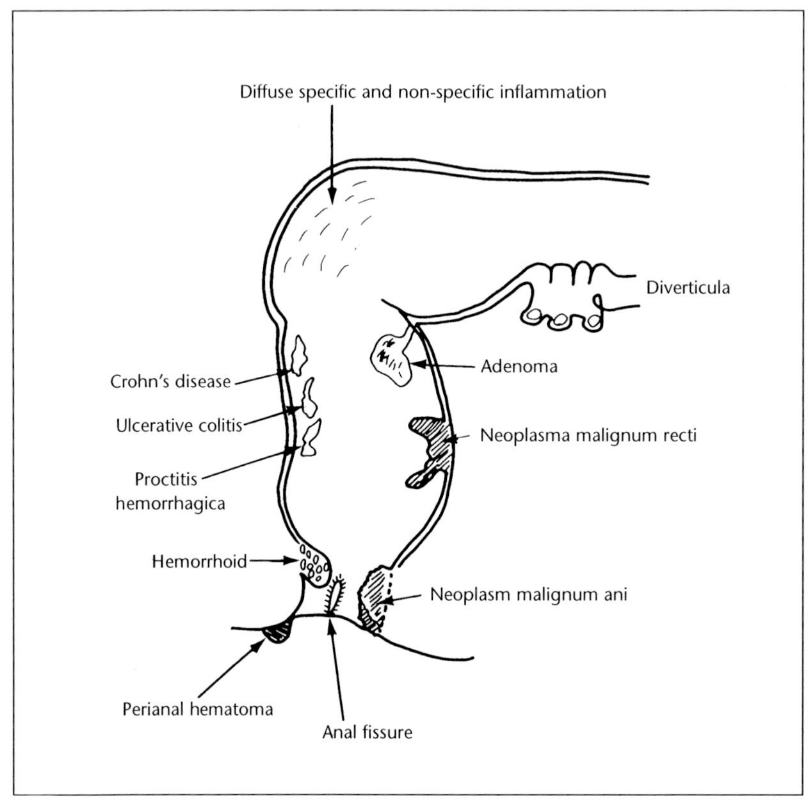

Fig 2.1. *Common reasons for perianal bleeding. By far the most common reason is hemorrhoids.*

Pruritus ani

Secondary
Idiopathic

In contrast, the following symptoms or types of pain suggest disorder in the anal canal itself. Pain is often sharp and severe in the anal canal because of its somatic innervation. It is of differential diagnostic significance if the pain is continuous or corresponds to defecation since pain that is not related to defecation is unlikely to arise from the anal canal.

Tenesmus

Burning sensation in the anal canal accompanied by sphincter spasm and an urge for defecation and a feeling of incomplete emptying.

Malignant growths
Inflammatory processes
Painful sphincter (sharp rhythmic pain that occur at the end of defecation)
Anal fissures that extend up into the anal canal

Sharp continuous pain

Abscess
Tumor
Thrombosed internal hemorrhoid
Inflamed hypertrophied anal papilla

Dull aching exacerbated by straining

Intrasphincteric abscess

Pain arising from the anal canal may be perceived as coming from the sacral region. In such situations information concerning pain relationship to defecation will often have diagnostic value. Proctalgia fugax (spasm of the levator ani muscle) is often misinterpreted as being caused by hemorrhoids or anal fissures. Pain localized to the coccyx rarely has an anal origin. It is most often idiopathic or a result of previous trauma. It may also be caused by infection in a presacral cyst.

DISCHARGE (SOILING)

Secretion of mucus which is produced by the goblet cells of the rectum may be seen in the stool in many situations. It may be an early sign of villous adenoma in the rectum or indicate an early colitis. The use of some suppositories may stimulate mucus secretion. If the secretion is, however, tinged with blood this is often a sign of an inflammatory or malignant process.

Soiling may be seen with various disorders of the anal canal and rectum. The soiling may be purulent or fecal. It may be due to a mild sphincteric incontinence, but may also arise from the patient wiping toilet paper high up in the anal canal causing irritation of the mucosa and a subsequent reflex atoni of the sphincter muscles. Soiling is not infrequently seen with malignant lesions of the anal canal. Fecal soiling is also common after construction of an ileo-anal anastomosis because of the relative atoni of the sphincter muscles or because of infection in the reservoir.

ALTERED BOWEL HABITS

Many patients with anal disorders will complain of irregular bowel habit or a change in bowel habit. The interpretation of this is not yet clear since it is difficult to define altered bowel habit as the normal pattern varies greatly from patient to patient and in the individual patient. The causes of altered bowel habits are often multiple and one should investigate the patient thoroughly and give a clear definition of what is meant by altered bowel habit. As regards constipation, for example, one might understand hard stool, infrequent stool or difficulties with rectal emptying.

PROLAPSE SYMPTOMS

Patients with disorders of the anorectum often complain of prolapse symptoms. Prolapse symptoms may occur in relation to defecation or may be unconnected with it. Symptoms of prolapse independent of defecation suggest that the cause is a hypertrophied anal papilla or full thickness prolapse. A prolapse may disappear spontaneously or may require digital replacement. Prolapse occuring with defecation that requires digital reduction is must likely caused by third-degree hemorrhoids (Fig. 2.2), a situation which must be clearly separated from full thickness prolapse. Other causes of prolapse are polyps in the rectum in children (juvenile polyps) or older persons (large villous adenomas of the rectum).

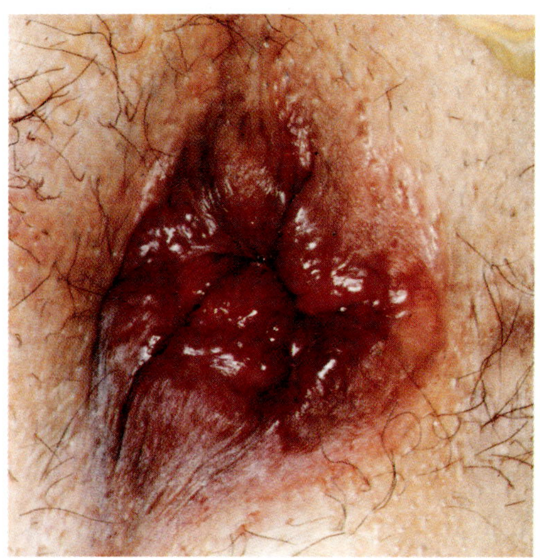

Fig 2.2. *Hemorrhoids in the third degree.*

SUPPLEMENTARY READING

Jensen, SL. Comparison of podophyllin application with simple surgical excision in clearance and recurrence of first episode perianal condylomata acuminata. Lancet 1985; i: 1146.

Jensen, SL. Treatment of first episodes of acute anal fissure: prospective randomized study of lignocaine ointment versus warm sitz baths plus bran. Br Med J 1986; 292: 1167.

Jensen, SL, Harling H. Prognosis after simple incision and drainage for a first episode acute pilonidal abscess. Br J Surg 1988; 75: 60.

Jensen SL, Nielsen OV. Cancer ani i Danmark 1943–1973. II. Sjældnere former. Ugeskr Læger 1982; 144: 856.

Jensen, SL, Hagen K, Shokouh-Amiri MH, Nielsen OV. Does an erroneous diagnosis of squamous cell carcinoma of the anal canal and anal margin at first physician visit influence prognosis. Dis Colon Rectum 1987; 30:345.

Jensen, SL, Harling H, Tange G, Shokouh-Amiri M, Nielsen OV. Maintenance bran therapy for prevention of symptoms after rubber band ligation of third-degree haemorrhoids. Acta Chir Scand 1989; 154: 395.

Nielsen OV, Jensen SL. Cancer ani i Danmark 1943–1973. I. Carcinoma planocellulare ani – diagnose, prognose og behandling. Ugeskr Læger 1982; 144: 851.

III. Examination and instrumentation

As discussed in chapter 2 much progress can be made in diagnosis by taking an accurate history; a correct diagnosis, however, can only be achieved with the help of a thorough objective examination supplemented by any necessary biopsies, X-ray examinations, and functional studies. Omitting anorectal examination in patients complaining of symptoms from the anorectal region must be regarded as professional negligence.

As with all other physical examinations the procedures should be carried out systematically. Examination should consist of inspection, palpation, digital rectal exploration, anoscopy, and rectoscopy. One should of course carry out a general objective examination of the patient if there is anything to indicate more than just a banal anal disorder.

One of the prerequisites for obtaining maximum information from the physical examination of this region is that the patient is secure and relaxed with the situation and is comfortably positioned. Patients are often nervous and tense in the examination situation. They should be well informed about the nature of the examination and should be continuously informed about what is being done during the examination. Cold instruments should not be used and instrumentation of the anal canal should be carried out with copious amounts of lubrication cream. If, in spite of these precautions, the examination cannot be carried out, for example, because of severe pain in the anal region the examination should be carried out under general or local anesthesia.

Another important precondition for a successful examination is the use of good illumination and appropriate instruments.

Anal examination should always be undertaken in the presence of a nurse.

POSITIONING OF THE PATIENT

Correct positioning of the patient is of great significance for both patient and doctor. It is important, to prevent back problems, that the examining doctor assumes a comfortable working position at the examination table. Correct working height is critical and a height-adjustable table is helpful in achieving this.

The lithotomy position

The gynecological or lithotomy positioning of the patient is nearly always used in Denmark for anorectal examinations (Fig. 3.1). The patient is positioned with the buttocks hanging a few centimeters beyond the edge of the table. The buttocks are usually slightly elevated and the legs held up in supports so that the hip joints are flexed (75°–80°) and slightly abducted (15°), and knees flexed (45°–50°) in the supports. The leg supports should be positioned at the correct height according to the patient's size and they should be angled so that there are no sharp edges to make the patient uncomfortable.

The lateral position

If the patient is only to be examined by inspection and palpation the lateral position

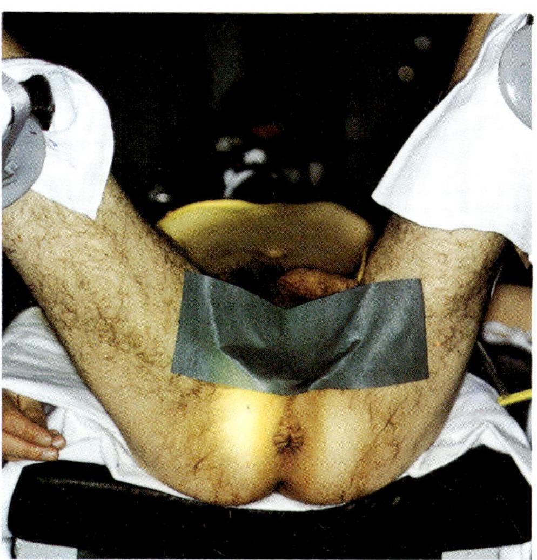

Fig 3.1. *The lithotomy position.*

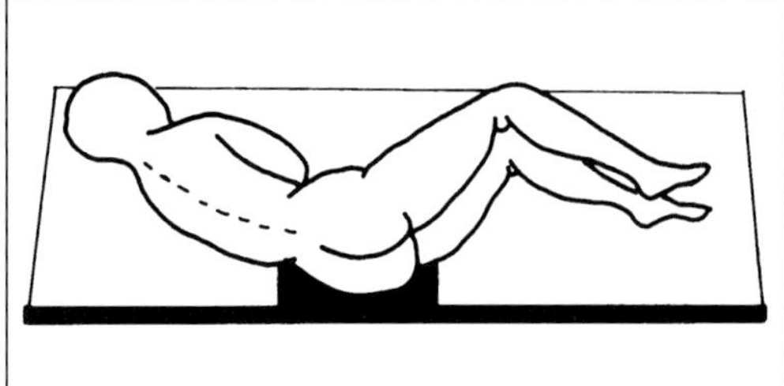

Fig 3.2. *Lateral position.*

(Sim's position) is often used. The patient is positioned in a left or right lateral position with flexed hips and knees, preferably with a cushion under the hips (Fig. 3.2).

The knee-elbow position

We have no experience with this position but it is used by some doctors and seems especially appropriate for anorectoscopic examinations but is not thought to be especially comfortable for the patient (Fig. 3.3).

INSTRUMENTS AND UTENSILS

Anoscope

There are many different types of anoscopes on the market both for single and multiple use with or without built in fiberoptic illumination. The majority of anoscopes are built for direct vision but some are made for lateral viewing. Anoscopes are constructed in different lengths and thicknesses and with straight or bevelled ends. We always use a Welch-Allyn-type anoscope which is 5 cm long and has a diameter of 2.5 cm and a bevelled end (Fig. 3.4). The

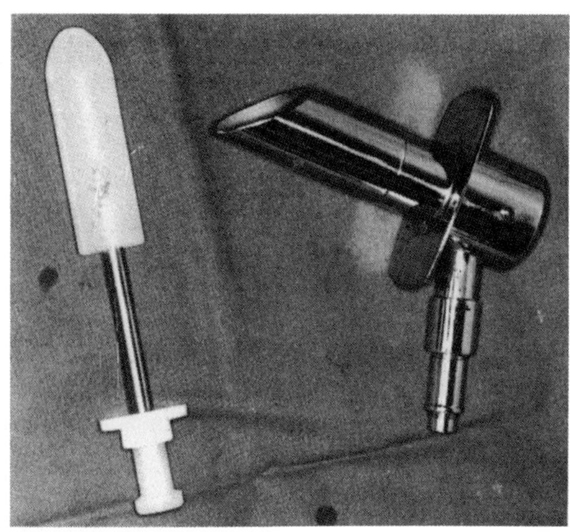

Fig 3.4. *Anoscope.*

reason for using a bevelled anoscope is that it is suitable for various therapeutic procedures, such as ligation of internal hemorrhoids, while anoscopes with straight ends are not. An anoscope consists of a tube and an introducer which is removed after the anoscope has been inserted into the anal canal. Fiberoptic anoscopes can be connected to a light source via a cable. Anoscopes without light sources provide less effective viewing conditions and should not be used. The light may come from a lamp behind the examiner or from a source which a nurse holds for the examiner. This form of illumination is inferior to instruments with built-in illumination.

Indications for anoscopy are obvious. All patients with anorectal symptoms should be examined anoscopically, including patients with metastases from malignant melanoma, where the tumor's primary site in unknown. Patients with ulcerative colitis or Crohn's disease should also undergo anoscopic examination.

Rigid Proctosigmoidoscopy

Rigid proctosigmoidoscopes are available as single-or multiple-use instruments. They are constructed along the same lines as anoscopes but all rectoscopes have built-in light sources. Rigid proctosigmoidoscopes also consist of an outer tube with an introducer (Fig. 3.5). The tube of a reusable scope is made of metal while disposable, single-use instruments are plastic.

Fig 3.3. *Knee-elbow position.*

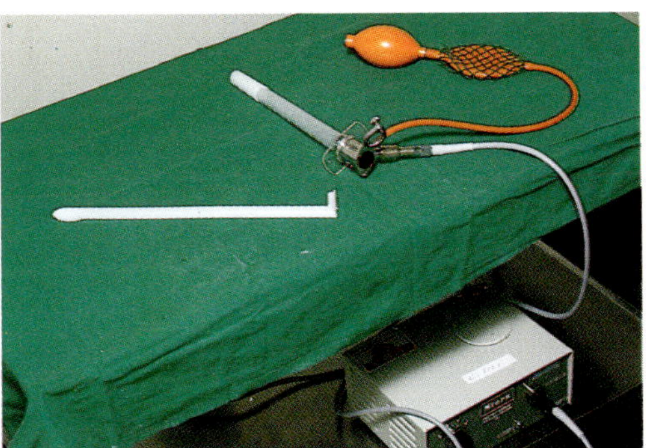

Fig 3.5. *Sigmoidoscope consisting of introducer, scope and balloon.*

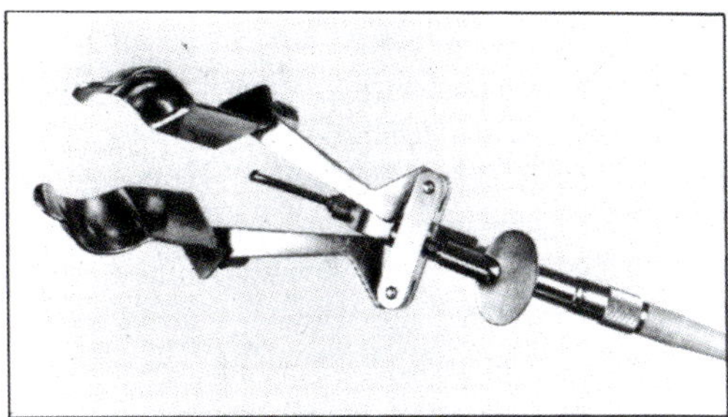

Fig 3.6. *Anal speculum with in-built light source.*

Single-use scopes are usually 25 cm long as opposed to multiple-use instruments which are 30 cm long. The diameters vary from 2 to 2.5 cm. In contrast to anoscopes, they have a glass viewing window which can be closed during examination so that air can be insufflated into the system without escaping. Air is introduced by means of a balloon attached to the scope in order to distend the bowel (Fig. 3.5).

Different proctosigmoidoscopes are available for diagnostic and therapeutic purposes. Operation scopes, which have a wider caliber than the usual diagnostic scopes, are used for the treatment of lesions in the rectum. It is possible, however, to remove rectal polyps with the usual kind of scope using a wire loop, biopsy forceps or a diathermy coagulator. These instruments can also be used with an anoscope.

The pediatric scope has a length of 20 cm and a diameter of 1.2 cm and is available with the same accessories and apparatus as the adult scope. The pediatric rectoscope is used in precisely the same way as the adult scope. The only difference is its size.

Anal specular

Various types of anal specula are available which are used for different operations in the anal canal, for example fissures, fistulae, and superficial tumors. They can only be used in patients under full general anesthesia or after effective local anesthesia of the area. We use Park's speculum with a built-in light source (Fig. 3.6).

Biopsy instruments

An ordinary scalpel or rectal biopsy forceps can be used for taking biopsies from the anal canal or the perianal region. Biopsies from the rectum are usually carried out with the type of forceps shown in Fig. 3.7, a diathermy loop or a special instrument for suction biopsies. When taking biopsies from the lower part of the rectum (the distal 8–10 cm) one can apply an elastic band with McGivney's apparatus to the biopsy site to prevent bleeding, or one can electrocoagulate bleeding sites resulting from biopsies higher up in the rectum.

Swabs and suction

It is important that the field of vision is kept clean during ano- or rectoscopic examination.

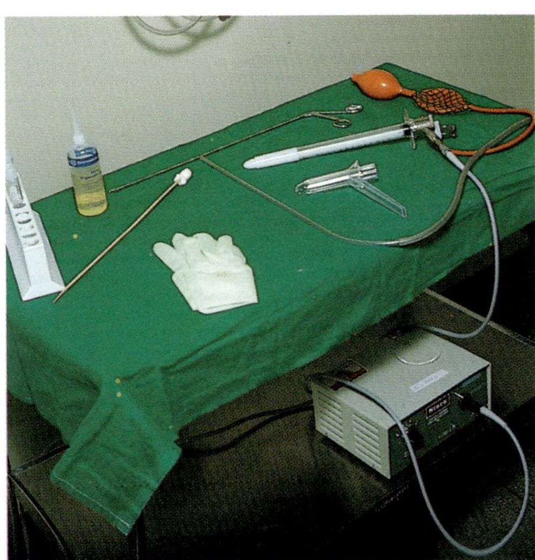

Fig 3.7. *Equipment for sigmoidoscopy: Sigmoidoscope with illuminator, lubricant cream, anoscope, biopsy forceps, gloves, gauze swabs, suction and formaline.*

Often the patient's bowel has not been sufficiently cleansed and it is necessary to swab or suck the field clean. Swabbing can be done with prefabricated gauze swabs mounted on long applicators (Fig. 3.7). They can be introduced via the ano- or rectoscope. The swabs may also be applied with special swab forceps. Instead of swabs one may use suction to clean the mucosa of feces or secretions. Special suction apparatus is available for this purpose. Sometimes cleansing has been so inadequate that swabbing is insufficient to clear the lumen and only suction can cleanse the area. One should, therefore, have both swabs and suction prepared for every anorectoscopy.

Lubrication cream

Cream or jelly is used to lubricate gloves and instruments so that introduction of these into the anal canal is as comfortable as possible for the patient. Lubrication cream contains sodium carboxymethylcellulose which is pharmacologically inactive. This substance can be easily rinsed from the instruments with water.

EXAMINATIONS
Anorectoscopy
Preparation of the patient

Patients to be examined anoscopically or rectoscopically without anesthesia need no bowel preparation. If the patient is to be examined under general anesthesia he should be fasting, prepared for anesthesia, and should have had his bowels emptied. Emptying of the anorectum can be achieved with rectal laxatives based on two principles. The first is softening of the feces in the rectum with substances that stimulate mucosal secretion and the second is stimulation of the defecation reflex. This reflex is initiated by stimulation of the nerve endings in the rectum which can be achieved by distension of the ampulla or by irritation of the mucosa with hypertonic fluids or contact laxative. Lubricating and softening enemas should be retained in the bowel for a few hours. Mucosa-stimulating preparations usually work after 5–20 minutes and are released together with the feces. We normally recommend one or two Microlax® 1

and 2 hours before examination. This substance works after 5–20 minutes. It is given rectally from 5 ml tubes. We advise against the use of oral laxatives as preparation for anorectoscopy.

Patients with anal incontinence should have their first anorectoscopy without cleansing or evidence of fecal impaction causing anal incontinence might be overlooked.

Inspection

Every examination of the anorectal region should be preceded by a thorough inspection of the perianal region and the anal opening. Inspection of the perianal region should be done with the patient correctly positioned and with good illumination. Inspection of the perianal region is often carried out very superficially or not at all during rectal exploration, for example, with the patient lying on his back in bed with inadequate lighting. Inspection should be undertaken with the patient in the lithotomy position and should involve examination of the perianal region by gently retracting the buttocks. The anal opening and anal canal should be examined by placing a gauze swab at each side of the anus and retracting well at each side (Fig. 3.8). The presence of skin changes (pruritus ani), tumors, condylomata, skin tags, external hemorrhoids, anal fissures, etc. should be noted. Inspection of the

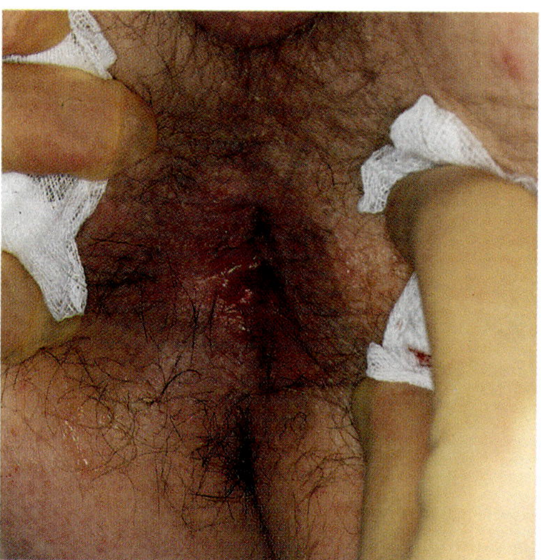

Fig 3.8. *Inspection of the anus and anal margin by means of pulling with swabs on each side of the anus – finding the superficial fissure and the condyloma.*

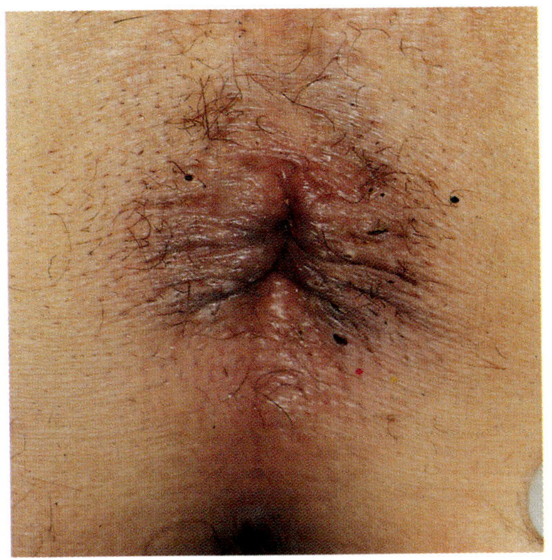

Fig. 3.9. *The normal appearance of the anus.*

buttock cleft is best carried out with the patient in the Sim's position or lying prone. Here, one is especially looking for the presence of a pilonidal sinus. Fig. 3.9 shows the normal appearance of the anus.

Palpation

A thorough palpation of the anal region can give significant information of a mass. Is it tender, is there fluctuation, and can one express pus from the lesion? Is there induration corresponding to the course of a fistula.

Rectal examination

Rectal examination commences with lubricating the anal opening and the index finger with lubrication cream. The finger is then carefully inserted into the anal canal (Fig. 3.10). Simultaneously one buttock is gently retracted to the side with the contralateral hand. During rectal examination, the intersphincteric groove and the anorectal ring should be palpated. The tone of the sphincter apparatus and puborectalis should also be investigated by asking the patient to squeeze around the examining finger. If the tone is diminished, normal or increased one looks for any tenderness or pain during palpation as well as searching for indurated areas or masses in the anal canal. It is possible to palpate chronic fissures, non-specific hypertrophied

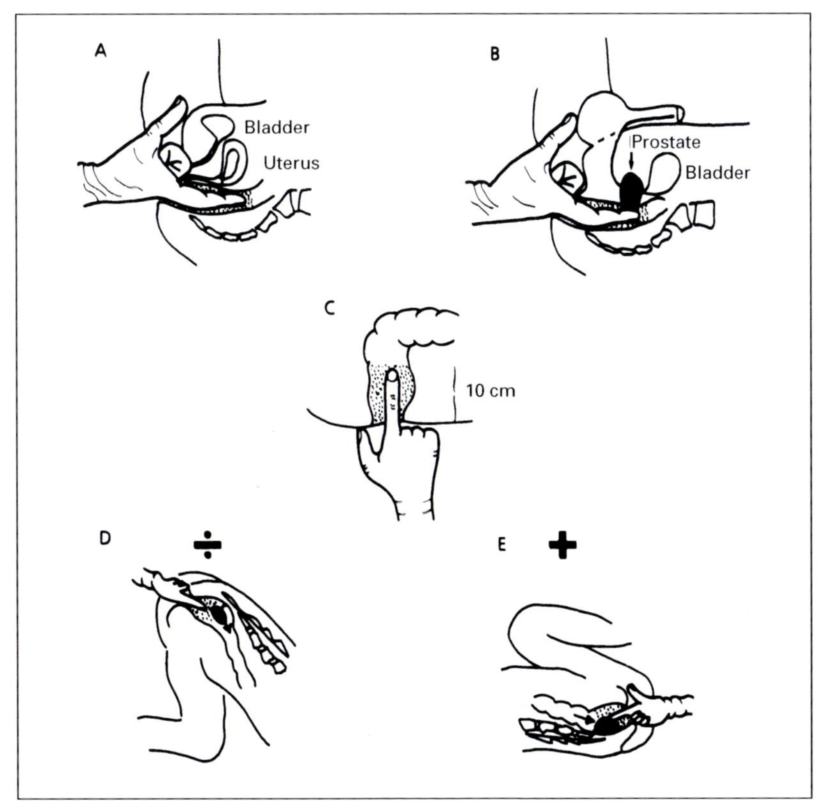

Fig 3.10. *The principles of rectal exploration.*

anal papillae, polyps, the induration at the site of internal fistula opening and malignant tumors. Abscesses can also sometimes be palpated during rectal examination. An ischiorectal abscess will often bulge into the anal canal. Internal hemorrhoids cannot normally be palpated.

In male patients the prostate (Fig. 3.10) and the seminal vesicles can be recognized while in women the cervix of the uterus is noticed. The body of the uterus (Fig. 3.10) can be felt if it is retroflexed. It is necessary to distinguish the uterus from a rectal cancer.

The rectum is palpated as high up as one can reach (Fig. 3.10), taking note of the rectal walls and any content it may hold. The normal rectal mucosa feels soft. Rectal cancer is easily recognizable by induration and umbilication; a pedunculated polyp can, however, be difficult to find. If a rectal cancer is discovered its relationship to the anorectal ring and the uterine cervix or prostate should be described. It is also important to gain an impression of the tumor's mobility and invasion of surrounding structures. When the rectal examination has been completed the exploring finger should be examined. If there is blood, pus or mucus on the glove it might indicate the presence of a tumor or an inflammatory process (Fig. 3.11). The inguinal lymph nodes should be palpated before or after the rectal examination. If enlarged lymph nodes are found their consistency, tenderness, and relationship to

the skin should be described. It is especially important to carry out a thorough rectal examination as part of the general physical examination of a patient even if they have no anorectal symptoms.

The authors have no doubt that routine rectal examination is very useful if it is carried out correctly. Unfortunately, however, the examining doctor often fails to do a rectal examination. It should be emphasized that many patients with anorectal cancer have had their cancer overlooked because rectal examination has been omitted during admission for some other disease. Every doctor should be able to carry out and interpret a rectal examination correctly. The diagnosis of tumors at rectal examination is best carried out, in our opinion, with the patient in the lithotomy position because of the reasons shown in Fig. 3.10.

Anoscopy

Anoscopy is as a rule carried out in Denmark with the patient the lithotomy position. Before the anoscope is introduced a rectal examination should be carried out in order to diagnose painful conditions such as anal fissures, sphincter spasm, and anal tumors. The anoscope is introduced with the obturator in place (Fig. 3.12) and well lubricated. At the same time one of the buttocks is retracted laterally to give a clear view of the anal opening. The anoscope is carefully introduced through the anal opening with the

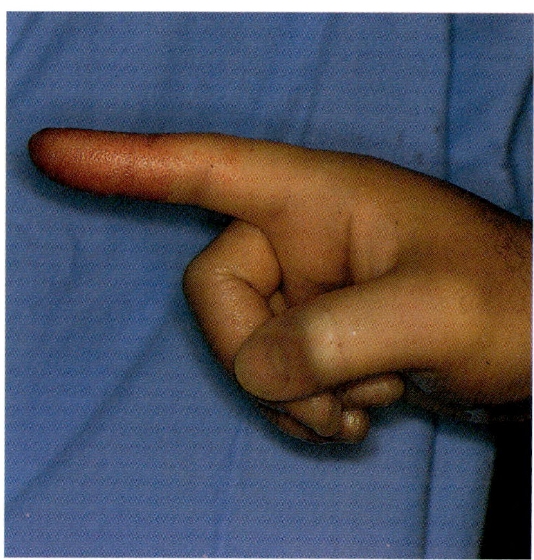

Fig 3.11. *Blood on the finger after anal exploration.*

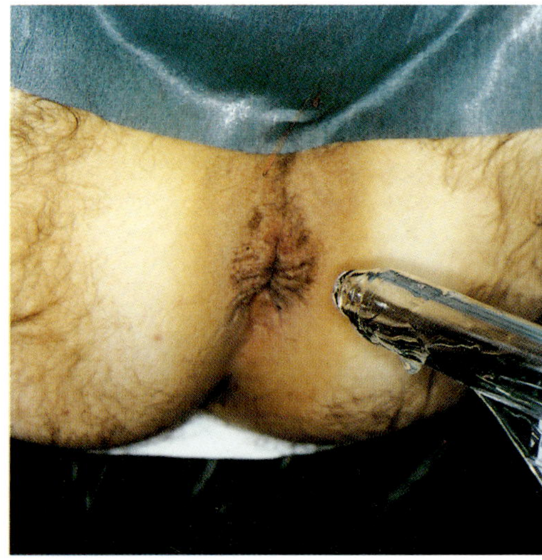

Fig 3.12. *Anoscope with obturator in place.*

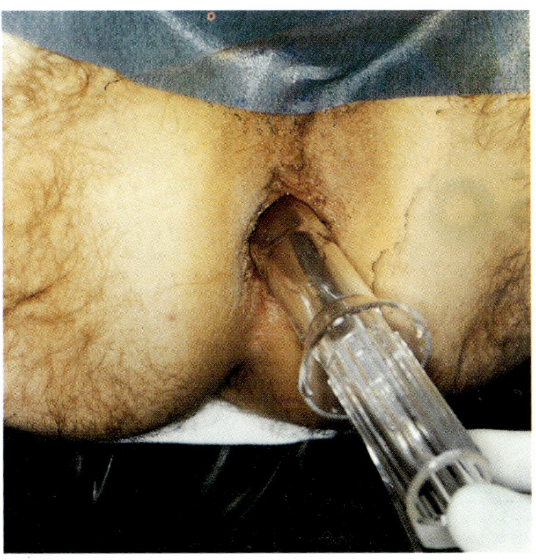

Fig 3.13. *Introduction of anoscope in the direction of the umbilicus.*

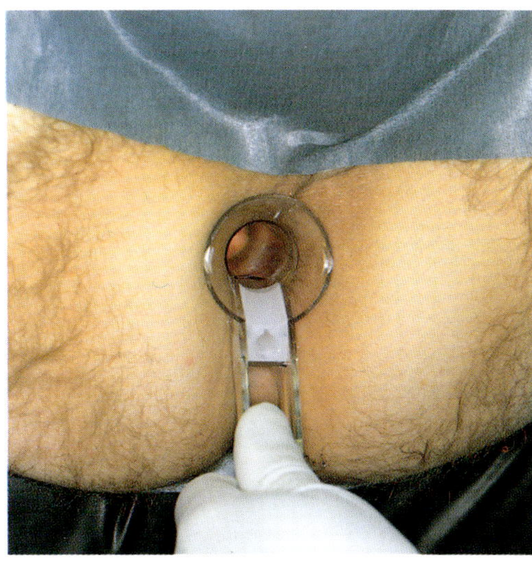

Fig 3.15. *The anoscope is in place.*

obturator in place and directed towards the umbilicus which corresponds to the axis of the anal canal (Fig. 3.13). The sphincter muscles gradually relax and the anoscope is easily introduced to its full length. The instrument is now directed posteriorly to obviate pressure on the prostate in men (Fig. 3.14). When the anoscope has been fully introduced it is supported in the left hand (Fig. 3.15) so that the examiner's right hand is free for other purposes, such as suction or swabbing the anal canal. The mucosa is studied carefully (Fig. 3.16) and biopsies may be taken or treatment instituted through the scope. At the end of the procedure the anoscope is removed from the anal canal. It is a good idea to ask the patient to strain while the anoscope is being removed in order to see whether a mucosal prolapse or hemorrhoids present themselves and to see whether these conditions reduce themselves spontaneously when straining ceases. During retraction of the scope the appearance of the anal crypts is noted. If the anoscope needs to be reintroduced the obturator should always be used even if the anoscope has not yet been fully retracted, otherwise unpleasant pain can be caused.

Proctosigmoidoscopy

Proctosigmoidoscopy should always be preceded by rectal examination. After correct positioning of the patient the scope is introduced blindly through the anus with the obturator in place. The obturator is removed as soon as the scope has traversed the anal canal and is introduced towards the umbilicus since the anus runs diagonally towards the umbilicus. In order to achieve a complete view of the ampulla the sigmoidoscope is directed posteriorly and is then gently introduced proximally and ventrally, possibly during gentle air insufflation using the balloon. This should not be performed violently as it causes unnecessary pressure and discomfort

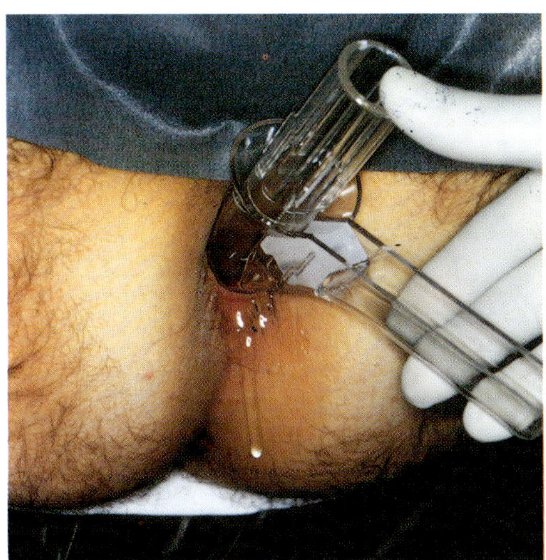

Fig 3.14. *The anoscope is moved backwards away from the prostate.*

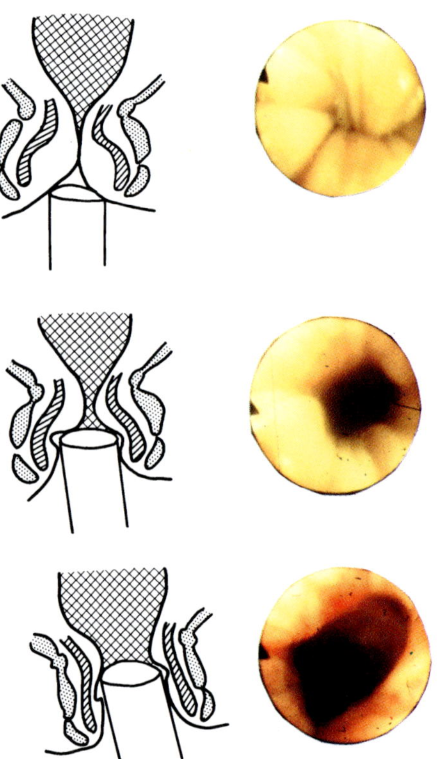

Fig 3.16. *Normal appearance of the anal canal seen through the scope.*

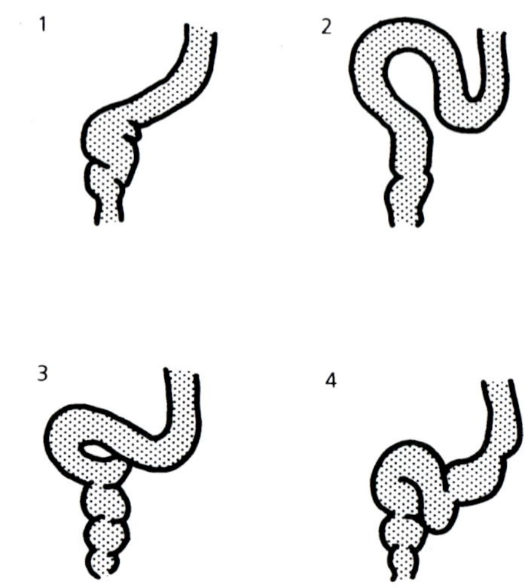

Fig 3.17. *Different angles between the rectum and the sigmoid.*

for the patient. The sigmoid colon usually begins 14–15 cm from the anus but there is a great individual variation. The angle between the rectum and the sigmoid colon (Fig. 3.17) varies from patient to patient. If the angle is too acute it may be impossible to carry out a rectoscopy even for the most experienced examiner. One should not force the scope through the angle because this carries a risk of perforation and causes unnecessary pain for the patient. A flexible sigmoidoscope should be used instead. Fig. 3.18 shows a schematic representation of the different steps taken during rectoscopy with photographs of the mucosa at the corresponding levels. If the patient's bowel has been insufficiently cleansed the mucosa can be sucked clean. If one finds a stenosis that cannot be passed with a normal rectoscope one should attempt to pass it with a pediatric scope.

Rigid proctosigmoidoscopy is, however, gradually being replaced by flexible fiberoptic sigmoidoscopy. The flexible fiberoptic sigmoi-

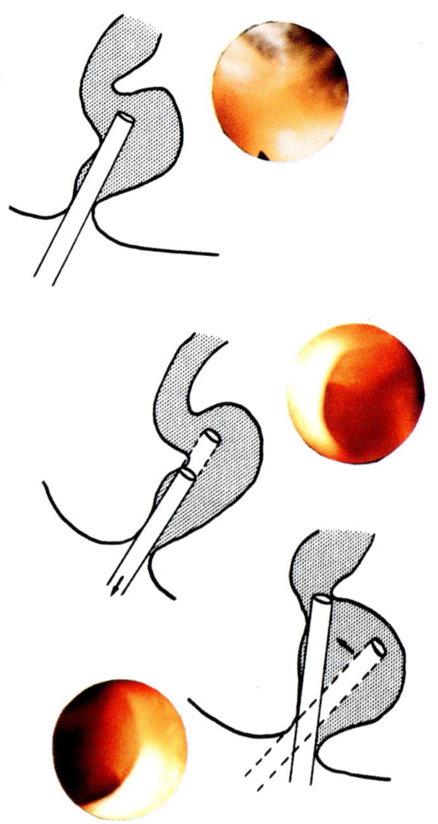

Fig 3.18. *Schematic representation of sigmoidoscopy.*

doscope is usually 60-65 cm long and because of the increased length, more pathological lesions (tumors, polyps and other diseases of the colon) can be found. The examination is usually carried out with the patient in the lateral Sim's position.

Indications

Anorectoscopy should be performed if there has been macroscopically visible bleeding from the anus or if there has been a positive test for blood in the feces. Fresh, bright red blood usually arises from the distal part of the colon or rectum and most often from hemorrhoids. It may also be due to a polyp, malignant tumors or inflammatory conditions. Other indications for anorectoscopy are changes in bowel habit, the feeling of incomplete emptying of the rectum, prolapsing mucosal fold, tumors, and changed appearance of the feces.

As mentioned above, the proctosigmoidoscopy with the rigid scope is usually replaced by the flexible fiberoptic sigmoidoscope today.

Contraindications

Endoscopic examination is contraindicated in patients who cannot cooperate as well as in patients with acute inflammatory bowel disease, such as fulminant ulcerative colitis, toxic megacolon, and acute diverticulitis.

Complications

Complications following proctosigmoidoscopy are extremely rare. A figure of one per 10,000–20,000 is given. The most common problems are perforations or bleeding after the taking of biopsies.

X-ray examinations

The use of conventional X-ray examination is of no great practical significance as a diagnostic tool in patients with anal disorders but a radiological colon examination is often ordered for patients over a certain age with rectal bleeding in order to exclude a colonic tumor. Some departments with a special interest in proctology, however, increasingly use functional X-ray examination for the investigation of special functional anal disorders.

Radiological examination of the colon (barium enema)

The anal columns of Morgagny can be demonstrated in a well executed double contrast examination of the colon. The examination can give information about pathological processes and about the rectum's configuration and the impressions made by the puborectalis muscles. This type of examination is, however, not routinely used for this purpose.

CT scanning

The muscle layers of the anal region can be differentiated from the surrounding fatty tissue at CT scanning. It is possible to demonstrate the levator ani muscles and the ischiorectal fossa (Fig. 3.19). The method can be used to assess the extent of pathological processes especially tumors which have spread to the regional lymph nodes.

NMR scanning

Nuclear magnetic resonance (NMR) is a relatively new form of radiological investigation that can be used to give information on both the anatomical and the physiological-biochemical relationships of the inner organs. Experience with NMR scanning in proctology is limited but so far the results in patients with anorectal tumors have been rather disappointing. NMR scanning makes use of the fact that the organism's different atoms function as magnets. When, for example, the anal region's different structures are subjected to a powerful external magnetic field every atom will position itself according to this external magnetic

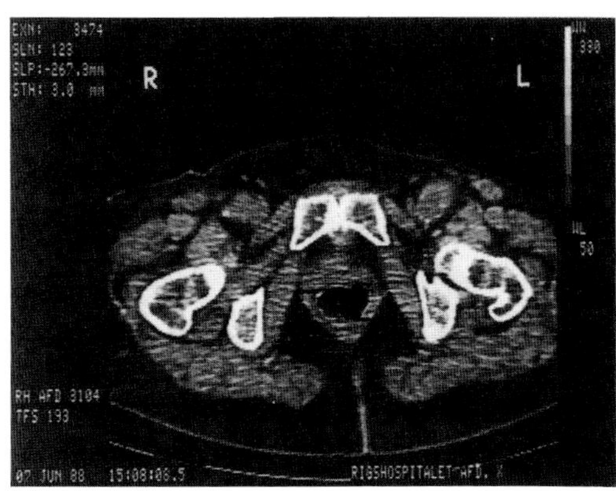

Fig 3.19. *CT of the anorectal region.*

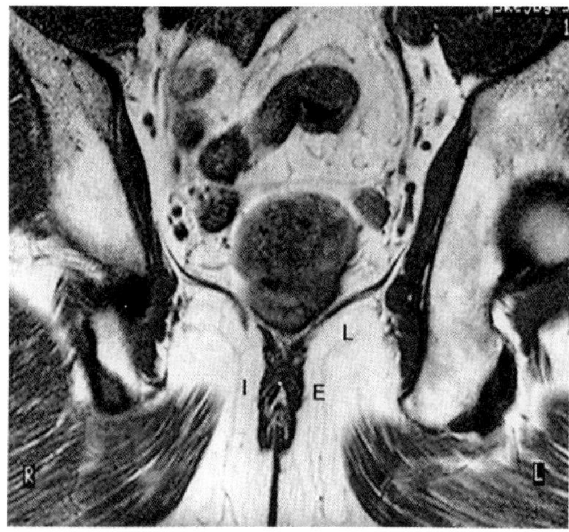

Fig 3.20. *MR section of the anal region.*

field. If one hereafter sends radio waves into the region some of the atoms (magnets) will absorb energy from the electromagnetic radiowaves and thus be rotated. If one disconnects the radio wave the magnets give off the energy they have absorbed. This energy can be recorded and a computer can construct a picture. This technique is still undergoing massive development and for detailed information the reader is referred to textbooks on NMR technology. In proctology NMR scanning will first and foremost demonstrate its worth in the investigation of complicated anatomical relationships. Fig. 3.20 clearly shows the ischiorectal fossa and its boundaries such as the levator ani and the external anal sphincter.

Defecography

This examination is used especially for the investigation of rectal prolapse and fecal incontinence. It is easy to carry out and gives good functional information. In spite of this, it is used clinically only in a few centers in Denmark. The method was first described in France in 1978.

Method. The patient presents himself without the bowel having been emptied. It is in fact contraindicated to use an enema, for example, since this can reduce the binding of the contrast materials to the bowel mucosa. The contrast material is made by diluting 150 ml of barium contrast (100%) with 400 ml of water. Potato flour (100 g) is stirred into the solution until a semisolid, white mass is produced. Using a pistol-like syringe with a tube introduced into the rectum, the rectum is filled with about 300 ml of contrast material. When the rectum has been filled, the tube is gradually removed from the anal canal while simultaneously injecting the contrast material, thereby clearly showing the anal canal during photographing. The patient sits upon a specially constructed chamber-pot stool so that pictures may be taken during defecation and straining. The examination is recorded on a videotape and takes about 15 minutes.

The examination gives the following information (Fig. 3.21–3.23):

1. The anorectal angle.
2. The level of the muscles of the pelvic floor at rest, during straining, and during defecation
3. The changes in the configuration of the rectum during defecation.

Ultrasound scanning

Ultrasound scanning for anal disorders has two purposes:

1. To show the possible hematogenous or lymphatic spread of a malignant lesion of the anus or to study regression or progression of tumors during treatment.
2. To demonstrate the extent of a tumor in the anus.

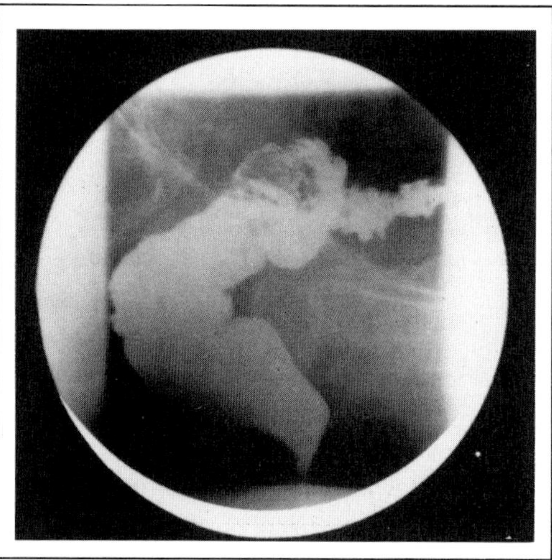

Fig 3.21. *Defecography: Normal anorectal angle.*

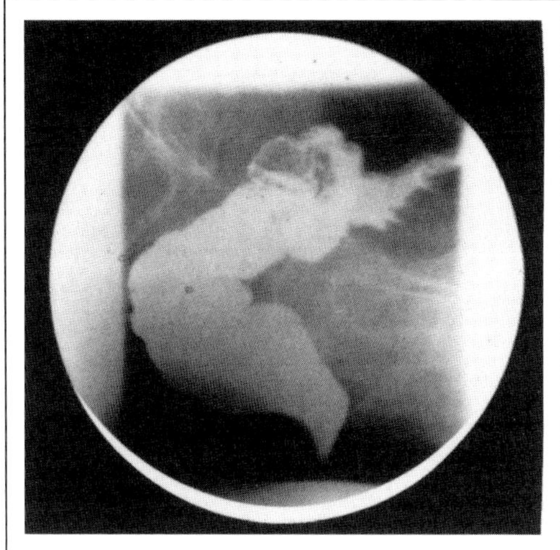

Fig 3.22. *Defecography: Normal appearance of the pelvis.*

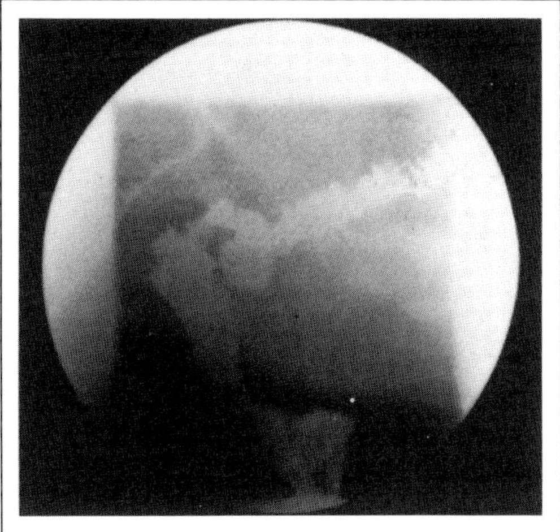

Fig 3.23. *Defecography: Internal polapse during defecation.*

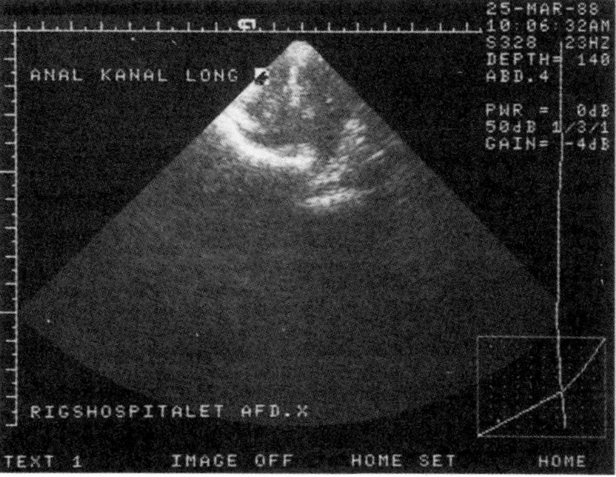

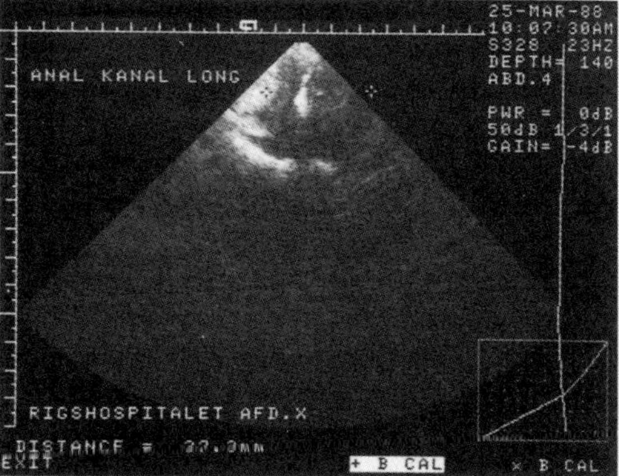

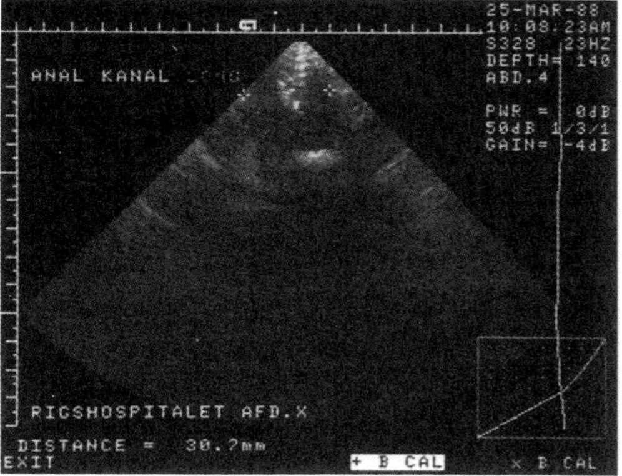

Fig 3.24. *Normal ultrasound scan of the anal region from the perianal region.*

In the latter case ultrasound scanning is carried out with the help of a transrectal scanner or by passing the ultrasound probe over the perineum and the perianal region (Fig. 3.24).

Microbiological investigations

Bacteriological investigations

In practice bacteriological investigations are only carried out during the treatment of anal abscesses and in the investigation of sexually transmitted diseases. For example, anal and rectal gonorrhea can be diagnosed by bacteriological cultures. It is important for a good culture result that the specimen is sent rapidly to the laboratory for investigation in a suitable substrate (for example Stuart's transport medium). The specimen is taken with a sterile charcoal-impregnated swab. The swabbing pen is introduced into the anal canal without contaminating it with feces. Swabbing from the rectal mucosa should also be done without fecal contamination.

Fungal investigation

Scrape and microscopy. The area from which fungus is to be isolated is cleaned with a sterile swab and moistened with 70% alchohol to remove dirt and any medications which would make microscopy difficult. The scrape should be taken from the active part of the lesion. If a blister is present its top is removed with scissors, dry lesions are scraped with a sterile curette, and the material placed on a slide on which a few drops of potassium hydroxide solution has been placed. The slide is then covered with a cover slip. Potassium hydroxide dissolves keratin which makes the specimen transparent. The slide is carefully warmed over a gas flame for a short time, then the specimen is pressed between two pieces of filter paper. The diagnosis of fungal infection is made by demonstrating the presence of yeasts or dermatophytic mycelia. The spores of yeast fungi lie in bunches while those of dermatophytes are found in chains.

Cultures. One can make cultures on different media. We use Sabouraud's agar (dextrose peptose agar). Media containing antibiotics, which inhibit the growth of bacteria, can also be used. The material is introduced into the medium under sterile conditions using a long needle. The inoculated medium is kept at room temperature and is examined regularly for two weeks. The appearance and color of the colonies are noted and studied microscopically. Yeast fungi will appear on the substrate within 2–5 days and dermatophytes within 5–30 days.

Examination for eggs

Enterobius vermicularis is found all over the world. It is most often children that are infected, but the whole family might be involved. The main symptom is severe pruritus ani.

Diagnosis. The eggs are best identified by applying a piece of transparent sticky tape to the perianal skin and then fixing the tape to a cover slip. The tape is studied microscopically to demonstrate the eggs adhering to it. The presence of eggs can be excluded only if multiple negative examinations have been done before the patient has bathed or defecated. The eggs may also be found under the nails.

Histological investigations

In our opinion all tissue removed from the anal region should be sent routinely for histological examination. Sometimes routine microscopy of an apparently benign hemorrhoid shows premalignant or malignant changes.

Taking biopsies

The site chosen for biopsies must be representative of the pathological process. Only macroscopically abnormal changes should be included in biopsies except for the tissue surrounding a tumor. Biopsy material can be obtained by various methods. A large incision biopsy with a scalpel is always welcomed by pathologists but it is often unnecessary in skin lesions of the perianal region. Concerning tumors that are difficult to diagnose, such as keratoacanthomata and squamous cell carcinoma, it is wise to take an incision biopsy instead of a punch biopsy. The latter can be adequate for perianal skin lesions. When an attempt has been made to remove a tumor *in toto* the surgeon should mark the tumor with ligatures so the pathologist is aware of the location of any possible residual tumor.

Fixation

The best fixative for lesions in the anal region is a neutral formalin buffer for ordinary routine examinations. The smaller the tissue fragments the better they are fixed. For skin lesions specimens of about 4 mm are optimal.

Staining

Hematoxylin and eosin staining are routine in most cases. Any need for special staining will be presented in the description of individual diseases.

SUPPLEMENTARY READING

Buie LA. Practical proctology, Springfield: Charles C. Thomas, 1960.
Gilbertson VA, Williams SE, Schuman L. The earlier detection of colonrectal cancers. I: Narjarian JS, Delaney JP, eds. Gastrointestinal surgery,

Chicago: Year Book Medical Publishers, 1979, 593–9.

Goligher JC. Surgery of the anus, rectum and colon, 5th ed. London: Balliere Tindall, 1984.

Marks G, Boggs HW, Castro AF, Gathright JB, Ray JE, Salvati E. Sigmoidoscopic examinations with rigid and flexible sigmoideoscopes in the surgeons office: a comparative prospective study of effectiveness in 1,012 cases. Dis Colon Rectum 1979; 22: 162.

Nesselrod JP. Clinical proctology. Philadephia: W.B. Saunders Co, 1964.

Paul PC, Palmer HH. Food theory and applications. New York: John Wiley and Sons, 1972.

Schrock TR. Examination of the anorectum, rigid sigmoideoscopy, flexible sigmoideoscopy, and diseases of the anorectum. I: Sleisinger MH, Fordtran JS, eds. Gastrointestinal disease, 4. ed. Philadelphia: Saunders, 1989.

Wiking N, Weissman GS, Gilbertsen VA, Winaver S, Sherlock P. A comparison of the 25–cm rigid proctosigmoideoscopy with the 65–cm flexible endoscope int he screening of patients with colorectal carcinoma. Cancer 1986, 57: 669.

IV. Congenital anal disorders

There are very few significant congenital anomalies of the anorectal region. In order to understand the background of the development of congenital anal anomalies the embryology of the region will be briefly revised.

ANAL EMBRYOLOGY

Until about the third week the embryo is flat and consists of two layers: the ectoderm and the endoderm. Thereafter the embryo develops into a three-layered plate which includes mesoderm. At the extreme proximal and distal ends the embryo still, however, consists of only the ectoderm and endoderm (Fig. 4.1) without interposed mesoderm. Proximally this two-layered point is called the buccopharyngeal membrane, which later becomes the mouth. Distally the point is called the cloacal membrane, which later becomes the anal opening or the anus. At about the fourth week the embryo changes to a cylindrical structure by curving in both the vertical and horizontal directions. During this curving process the yolk sac is pulled into the embryo and becomes the primitive gut that can be divided into the foregut, midgut and hindgut (Fig. 4.1). A small tubular diverticulum, the allantois, grows simultaneously into the body stalk, which is the communication between the embryo and the placenta. Next, a mesodermal partition, the urorectal septum, develops between the allantois and the hindgut and impinges upon the cloacal membrane. The hindgut is, thus, divided into an anterior part, a urogenital sinus and a posterior part, the anorectal canal. The anorectal septum finally fuses with the cloacal membrane and forms the perineum which divides the cloacal membrane into two parts, an upper urogenital membrane and a lower anal membrane (Fig. 4.2).

During the seventh and eighth weeks the anal membrane perforates and the anorectal canal becomes the rectum and anal canal. If perforation of the anal membrane does not occur or is incomplete a condition called imperforate anus results.

ANOMALIES
The perianal region

Severe anomalies rarely occur. On the other hand, minor anomalies are seen relatively frequently, for example, dermoid cysts near the scrotum, hemangiomata, skin tags, and papillary acanthomata. Hairy nevi are also occasionally seen and not infrequently congenital cysts, fistulae and sinuses, which can easily become infected. Pilonidal cysts or sinuses are frequently seen in buttock cleft.

The anal canal
Imperforate anus

The only congenital anomaly of practical importance is anal atresia. In this connection, it is important to ascertain whether one is dealing

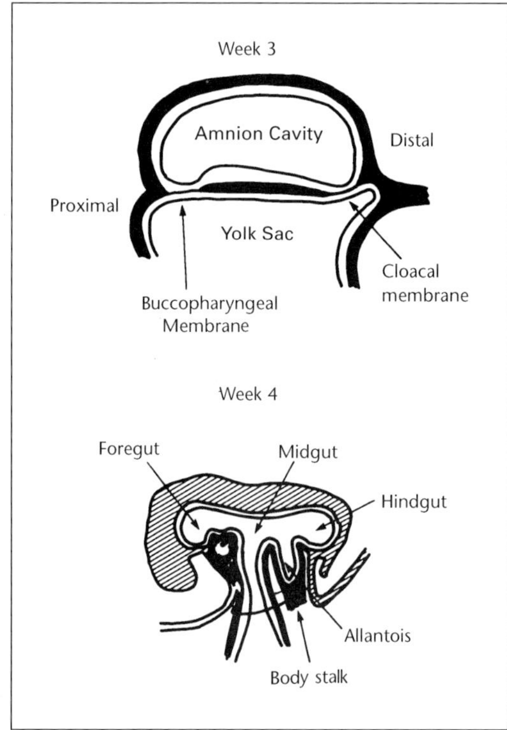

Fig. 4.1. *Longitudinal section through embryo weeks 3 and 4.*

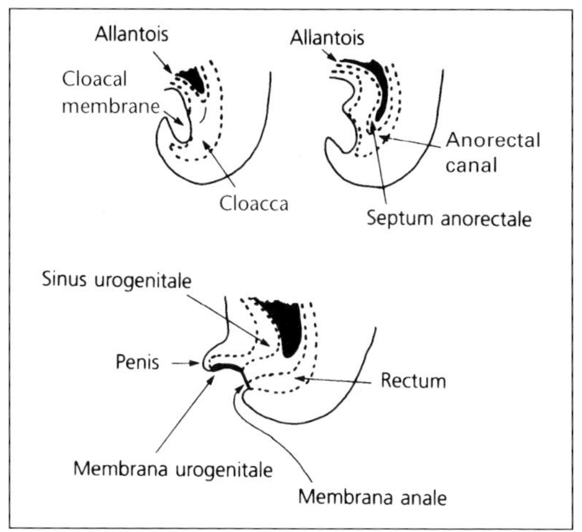

Fig. 4.2. *Sagittal sections of the lower part of the embryo.*

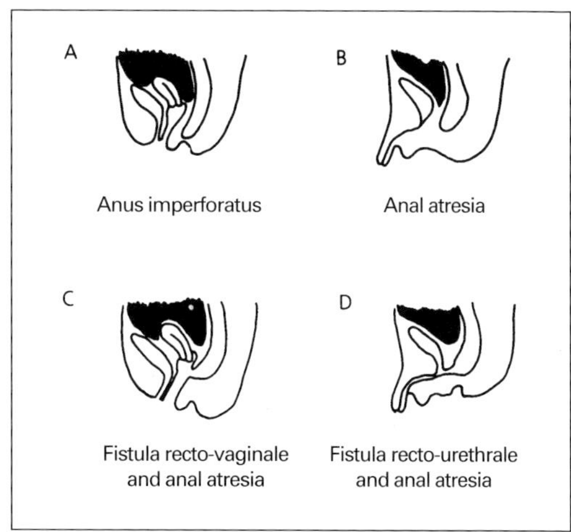

Fig. 4.3. *Various types of anorectal congenital abnormalites.*

with atresia with total obstruction, atresia with partial obstruction, or whether there is a fistula between the anorectum and the urinary tract, the genitalia or the perineum. This situation occurs rather infrequently in Denmark; there are about twenty cases annually pcr 100,000 births.

Patients with imperforate anus, as the condition is called, often have other congenital abnormalities. About 30% have anomalies of the genitalia and urinary tract, about 15% anomalies of the gastrointestinal canal, about 7% cardiac anomalies, and about 6% anomalies of the muscular skeletal system and CNS. One may divide the anomalies in various ways but one of the most commonly used classifications is that described by Ladd and Gross:

1. Stenosis of the distal rectum and anus (incomplete rupture of the anal membrane).
2. The membranous form of imperforate anus (persistent anal membrane).
3. Imperforate anus with the rectum ending blind at various distances from the perineum.
4. The anal canal and distal rectum creating a distal reservoir which is separate from a blindly closed rectum lying at a variable distance from the distal blind reservoir.

In most cases there is a fistula from the blind end of the rectum (type 3 and 4). In boys one might see anoperineal, anobulbar, rectovesical, or rectourethral fistulae. In girls anoperineal,

anovestibular, rectovestibular, rectovaginal, and rectocloacal fistulae may occur (Fig. 4.3).

Diagnosis of imperforate anus should be made during the physical examination that is undertaken immediately after birth. In most cases a little dimple can be found in the perineum at the normal site of the anus. In children with partial anal stenosis or a persistent anal membrane, the anal opening can have an almost normal appearance and can be located in the normal position. This anomaly may be overlooked and the diagnosis made later when attempting to measure the rectal temperature or when symptoms of intestinal obstruction develop at a later date. In the first instance the anal opening is often large enough to allow the passage of liquid stool. It is only after the stool becomes more solid that signs of commencing intestinal obstruction develop. It is important at the physical examination of children with imperforate anus to search thoroughly for the presence of accompanying anomalies in other organ systems. With agenesis of the sacrum, absence of innervation to the perineum and sphincter apparatus might occur and problems of continence will ensue.

Clinical diagnosis is verified by radiological examinations. The so-called Wangensteen–Rice invertogram is used (Fig. 4.4). The child is X-rayed while hanging with its head down. Gas in the blind rectal stump allows localization of the level of atresia. This gas pattern is normally

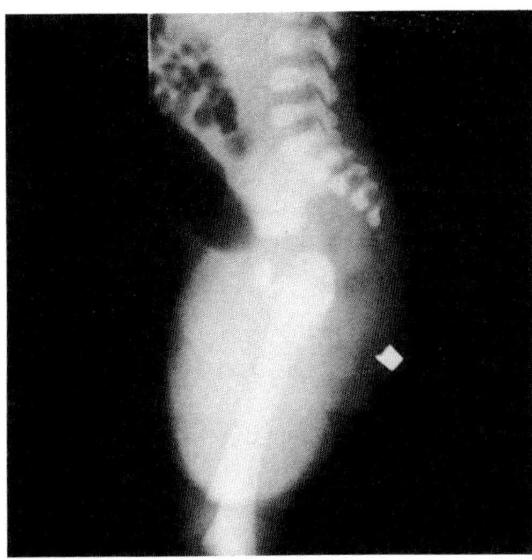

Fig. 4.4. *Wangensteen–Rice invertogram showing anal atresia.*

present about 12–24 h after birth. If there is an opening in the perineum the situation can be clarified by injecting contrast material through a narrow cannula. In most centers today the diagnosis is made simpler and more comfortable for the patient by ultrasound scanning. A fistula can be diagnosed by means of a urethrography or a vaginography. Treatment and investigation of these patients should take place in a pediatric surgical department but every doctor should be capable of making the diagnosis. Treatment of stenotic anomalies is carried out over a long period of time with repeated dilatations. On the other hand, obstruc-tive lesions with or without accompanying fistulae are treated surgically by specialist pediatric surgeons. Most children undergo the reconstructive procedure at the age of 1–2 years since the presence of a fistula will obviate defecation problems until then. If this is not the case a temporary colostomy is usually made. It is considered best that the children are operated upon at an early age to reduce the psychological trauma of the procedures and the accompanying hospitalization. A permanent colostomy should be avoided until the child has grown up. The prognosis depends on the presence of other anomalies and the nature of such anomalies as well as the type of anal atresia and the experience of the treating surgeon. The mortality of this condition is about 20% and about 75% of the survivors achieve a good result from the operative procedure. The remainder have incontinence that may require a permanent colostomy when the child is mature enough for this.

SUPPLEMENTARY READING

Adeyemi SD, Da Rocha-Afodu JT. Management of imperforate anus at the Lagos University Teaching Hospital, Nigeria: a review of ten years experience. Progr Paediatr Surg 1982; 15: 187.

Nielsen OH. Kongenitte sygdomme i fordøjelseskanalen. I: Hald T. Stadil F. Kirurgisk Kompendium København: Nyt Nordisk Forlag Arnold Busck, 1987: 883–920.

Swensson O, Donnellan WL. Imperforate anus. I: Swensson OR, ed. Pediatric surgery. New York: Appleton, Century, Crofts, 1969: 948.

V. Hemorrhoids

Definition

Hemorrhoids are normal anatomical structures. They are cushion-like thickenings above or below the mocucutaneous transition of the anal canal.

Anatomy

Hemorrhoids consist of a surface which may be mucosa or anoderm, and a stroma consisting of blood vessels, smooth muscle (the submucosal anal muscle), and connective tissue. Connective tissue fibers fix the hemorrhoids to the internal anal sphincter and the longitudinal muscle fibers between the external and internal anal sphincter. When this fixation becomes weakened or degenerate (commencing after 30 years of age) hemorrhoids may not only bulge into the anal canal but may also begin to prolapse through the anal canal. At the same time, the venous plexuses become stretched, resulting in the condition of clinical hemorrhoids.

Hemorrhoids may be internal or external. Internal hemorrhoids arise from the superior hemorrhoidal veins in the upper part of the anal canal above the dentate line. Such hemorrhoids are covered by either columnar epithelium or transitional epithelium. External hemorrhoids arise in the distal part of the anal canal in the inferior hemorrhoidal veins and are covered by anoderm or perianal skin. Often, a combination of internal and external hemorrhoids arises since there are anastomoses between these two venous plexuses.

The condition is called hemorrhoids when there is visible distension of the submucosal tissue in the anal canal. In most cases three hemorrhoids develop and the localization in the anal canal is almost always the same. Two of the hemorrhoids occur on the right side corresponding to seven o'clock and eleven o'clock (right posterior and right anterior) (Fig. 5.1); the third hemorrhoid occurs on the left side at the three o'clock position (left lateral) (Fig. 5.1).

Occasionally additional small hemorrhoids occur between the three major ones (accessory hemorrhoids). The size of the hemorrhoids is very variable. Initially they present as small cushion-like thickenings in the anal canal itself and which never prolapse outside the anus. These are classified as first degree hemorrhoids. Gradually the hemorrhoids enlarge and prolapse not only into the anal canal but appear outside the anal opening on defecation. If they continue to reduce themselves spontaneously they are defined as second degree hemorrhoids. Later the hemorrhoids prolapse more easily even at periods unrelated to defecation and require digital reduction. They are now called third degree hemorrhoids. Permanently prolapsed, skin-covered, irreducible hemorrhoids are regarded as fourth degree hemorrhoids. It is important to know these gradings since they are the basis for choosing the method of treatment.

EPIDEMIOLOGY

Clinical hemorrhoids are exceedingly common especially in the Western civilized world. It is difficult to assess their incidence but it is known to increases markedly with age; 50% of the population over 50 years of age have asymptomatic or symptomatic hemorrhoids. It

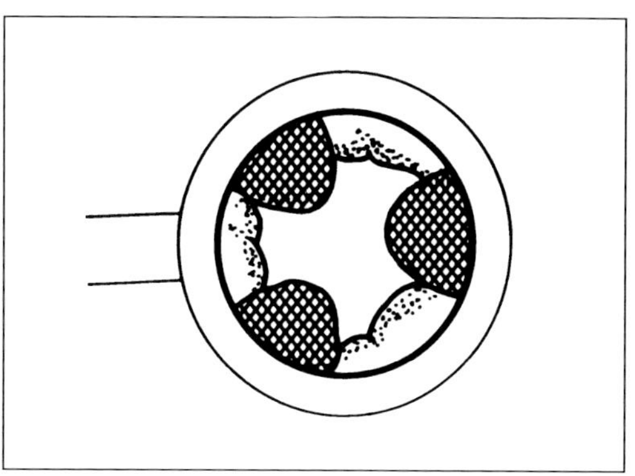

Fig. 5.1. *Location of internal hemorrhoids.*

is thus very common to find asymptomatic hemorrhoids during routine examination of the anorectal region.

Etiology

Generally, hemorrhoids first become symptomatic after repeated, forceful defecation of long duration which increases the venous pressure, distends the veins and causes loosening of the cushions from the underlying layers resulting in prolapse. Common causes of symptomatic hemorrhoids are therefore chronic constipation or diarrhea, as well as pregnancy, obesity and an inherited tendency. A large Danish study has recently shown that patients with symptomatic hemorrhoids have a significantly lower dietary fiber content than patients with asymptomatic hemorrhoids. The poor fiber content of the diet in Western society undoubtedly plays an important role in the high incidence of hemorrhoids.

Symptomatology

Hemorrhoids is one or the most common disorders of the anal region. The commonest symptom which brings the patient to the doctor is bleeding. This may be fresh streaks of blood on the stool or toilet paper or more massive bleeding which discolors the water in the toilet or the toilet bowl itself. Bleeding is most often bright red (arterial) because of the arteriovenous shunts in the hemorrhoidal cushions. On rare occasions the bleeding can gradually lead to iron deficiency anemia.

The other common symptom or complication is prolapse. As previously mentioned, prolapse occurs initially only on defecation and the hemorrhoids reduce spontaneously after defecation. As the disorder progresses, prolapse can occur at times unrelated to defecation (on coughing, sneezing, laughing or lifting heavy weights) and requires digital reduction. Permanent prolapse is seen with fourth degree hemorrhoids. As a result of prolapse the patient may also complain of pruritus ani because of copious mucus secretion from the prolapsed mucosa (soiling) which can cause constant irritation of the anal region.

Pain does not accompany uncomplicated hemorrhoids but is the result of prolapse, incarceration and thrombosis of the hemorrhoid. Pain occurring with uncomplicated hemorrhoids is usually due to an accompanying anal fissure.

Examination and diagnosis

Examination of patients with hemorrhoid symptoms should always include the following: inspection, rectal examination and anorectoscopy. The examination is carried out while the patient is in the lithotomy position.

Internal non-prolapsing hemorrhoids are not clearly visible on inspection of the anal region. In contrast, large third and fourth degree hemorrhoids are easily recognizable. Distally they are covered by skin and proximally by red mucosa and are usually separated by radial clefts. Hemorrhoids cannot usually be palpated on rectal examination; diagnosis is made on anoscopy. When the patient strains the hemorrhoids bulge into the anoscope which is gradually withdrawn to a position near the dentate line. Rectoscopy should always be carried out to eliminate the presence of malignant lesions or inflammatory processes higher up in the bowel.

X-ray examination of the colon or colonoscopy should only be carried out for specific indications (usually in patients over 40 years of age where there is a risk of a malignant lesion). It should, however, be done if the cause of bleeding is not revealed by anorectal examination.

Differential diagnosis

Since bleeding is the commonest symptom of hemorrhoids but it is also seen with many other anorectal disorders, it is important that such disorders are excluded by a thorough anorectoscopic examination. To exclude bleeding from higher up in the bowel (neoplasms, diverticulitis, inflammatory bowel disorders) it is, as mentioned, necessary in certain cases to supplement the investigations with an X-ray examination of the colon or colonoscopy. True rectal prolapse is easily distinguishable from prolapsed hemorrhoids as the whole rectal wall is involved and concentric mucosal folds are seen. Hypertrophic papillae, anal fissures, condylomata, and skin tags are easily distinguished from hemorrhoids by their characteristic appearances.

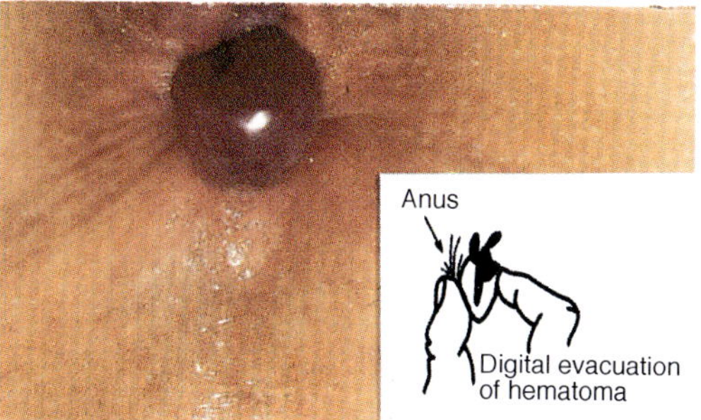

Fig. 5.2. *Perianal hematoma and treatment (digital evacuation of hematoma).*

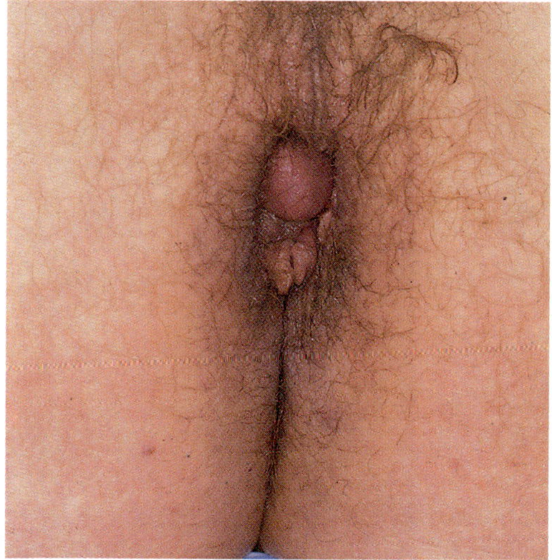

Fig. 5.3. *Incarcerated thrombosed hemorrhoid.*

It is important to distinguish a perianal hematoma (Fig. 5.2) from an incarcerated thrombosed internal hemorrhoid (Fig. 5.3) since the treatment is completely different.

Complications

Large prolapsed hemorrhoids may be trapped by a spastic sphincter with the development of incarceration, edema, and thrombosis. One or all three hemorrhoids may incarcerate. This is an extremely painful condition which makes sitting uncomfortable. On anal examination marked edema of the anal region is seen and hard thrombi may be found in the external parts of the prolapsed hemorrhoids. In certain cases the condition progresses to ulceration and necrosis.

Methods of treatment

Asymptomatic hemorrhoids found during routine examination require no treatment. Otherwise the treatment is as previously mentioned based upon the severity (degree) of the hemorrhoid.

Conservative treatment

Dietary fiber supplements. First degree hemorrhoids usually require no treatment but if they are symptomatic they can generally be handled conservatively. This involves regulating the patient's defecation habits, that is to say, giving instructions to avoid prolonged straining during defecation. The patient should be given a fiber-rich diet and, if necessary, daily fiber supplements together with copious fluid intake. The patient should be instructed to defecate as soon as the urge presents and to obviate lengthy visits to the toilet. If the hemorrhoid prolapses it should be reduced immediately after defecation. Fiber supplements should of course also be employed, even when other forms of treatment have been instituted, in order to prevent recurrence of the hemorrhoids.

Topical application of steroids and/or analgesics. Various suppositories and anal ointments are available for the treatment of hemorrhoid symptoms. The common denominator of these preparations is that they comprise either a steroid, a local anesthetic, an antibiotic, or a combination of these. However, the therapeutic effect of such suppositories and ointments on hemorrhoids has never been proved, and with prolonged use the ointments can even cause contact dermatitis or "steroid skin".

Sclerotherapy (injection treatment)

Sclerotherapy can be used for symptomatic resistent bleeding with first degree hemorrhoids. This treatment is rarely used in Denmark but is very popular in England. The purpose of the treatment is not, as with varices, to damage the intima of the veins with resultant thrombosis but rather to initiate an inflammation with resultant fibrosis in the submucosal tissue where the venous plexuses lie.

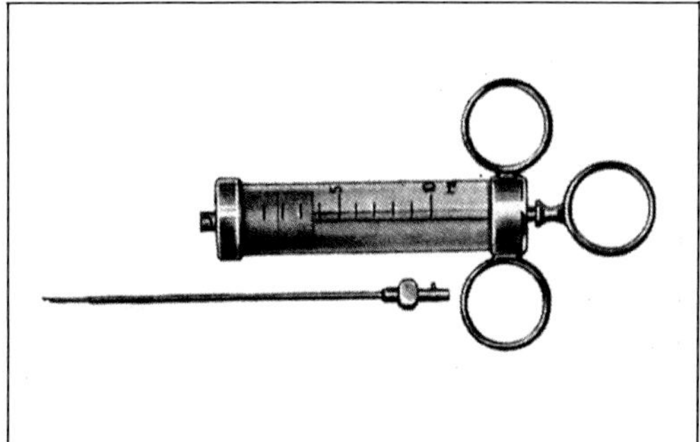

Fig. 5.4. *Syringe for injection of internal hemorrhoids.*

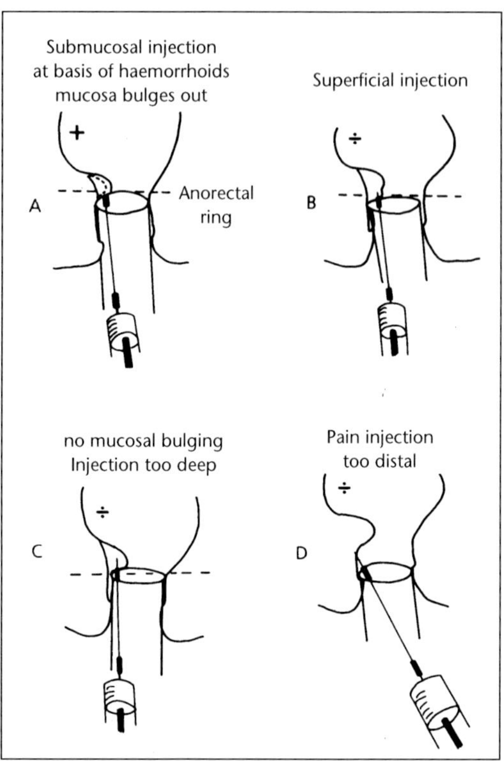

Fig. 5.5. *Injection treatment of internal hemorrhoids.*

This treatment can be done as an outpatient procedure without any preparation or local anesthesia. An anoscope with built-in illumination is used. A 10 ml syringe (Fig. 5.4) with three finger holes and a 7.5 cm long needle are employed (the needle has a stop about 2 cm from the point in order to mark its depth in relation to the surface of the mucosa) (Fig. 5.5A). A 2% solution of aethoxysclerol or 5% phenol in oil is used for the injection. 3–5 ml are used for each hemorrhoid with a maximum of 15 ml at each treatment. The patient lies in either the lithotomy or the lateral position. The anoscope is introduced into the proximal part of the anal canal where the anorectal ring is identified. The solution should not be injected into the actual hemorrhoid but into the submucous layer corresponding to its base just distal to the anorectal ring. This site is situated about 1–2 cm above the dentate line. The mucosa can be cleaned with a gauze swab prior to injection. The needle is introduced via the anoscope and obliquely through the surface of the mucosa. The above-mentioned stop can be used to assess the depth (Fig. 5.5A). Injection should be done slowly. About 0.5 ml is injected in the first instance. If this causes a white papilla on the mucosa (Fig. 5.5B) the injection is too superficial and the needle should be inserted more deeply. When the needle is correctly placed the mucosa slowly bulges during injection (Fig. 5.5). If the mucosa does not bulge the needle has been introduced too deeply and needs to be retracted (Fig. 5.5). If the patient complains of pain the injection has occurred too far distally (Fig. 5.5) and the needle should be removed and reintroduced more proximally in the rectum. The patient should feel nothing during treatment or at most a slight feeling of pressure in the anus.

This discomfort or pressure may last for up to 24 hours. Any bleeding from the puncture site can usually be stopped by compression with a gauze swab for a short time. All three hemorrhoids can be treated in the same session. If further injections are necessary they should not be carried out until 4 weeks after the initial treatment. One accurate injection treatment, however, is usually enough for bleeding first degree hemorrhoids. Injection treatment is contraindicated in the presence of inflammatory bowel disease, anal fissures or stenosis.

Complications. Incorrect intramucosal injections or the use of too large a dose of solution can cause necrosis and ulceration of the mucosa. Infection or temporary stenosis caused by edema is seen occasionally.

Rubber band ligation

Rubber band ligation of internal hemorrhoids is the most gentle and most frequently used procedure for second degree hemorrhoids and occasionally also for third degree hemorrhoids. Prolapsed hemorrhoids with a large external component do not usually disappear completely with rubber band ligation so it is often necessary in such cases to supplement the procedure with excision of skin tags.

Rubber band ligation cuts off the blood supply to the hemorrhoid which undergoes necrosis and disappears in the course of 10–14 days. This results in an ulcer of the mucosa which heals to leave scar tissue and fixation of the mucosa to the distal part of the rectum so that prolapse is prevented.

Rubber band ligation requires no special preparation of the patient and is done as an ambulant procedure without anesthesia. The apparatus used is illustrated in Fig. 5.6. The McGivney apparatus is simple in design. It consists of two metal cylinders mounted one inside the other. A handle activates the outer cylinder which by gliding over the inner cylinder pushes the rubber band off the inner cylinder where it has been previously mounted. Fig. 5.6 shows the apparatus with a rubber band mounted on the inner cylinder. The rubber band is mounted with the help of the cone.

An anoscope with built-in illumination is introduced into the rectum. The obturator is removed and the ligation apparatus introduced while an assistant holds the anoscope. The grasping forceps are introduced through the cylinder and the mucosa overlying the hemorrhoid (about 1–2 cm above the dentate line) is grasped and pulled into the cylinder, which is at the same time pushed towards the mucosa. If the patient has experienced pain when the mucosa was grasped by the forceps the rubber band should not be fired. If this is not the case the rubber band is fired and tissue the size of a cherry is ligated (Figs. 5.7 and 5.8). Severe pain after placement of the band is a sign of incorrect siting and the band should be removed. This may be a difficult task because of patient resistance (pain) but can be done most conveniently with a small pair of scissors or fine artery forceps. One to three hemorrhoids can be treated in the same session. Correct placement of the bands should be painless but a proportion of patients will experience a feeling of tightness in the anus together with a defecation urge. This mild discomfort may persist for 24–48 hours. If only one or two hemorrhoids are treated in the first session the next ligation treatment can be done 2–3 weeks later. With

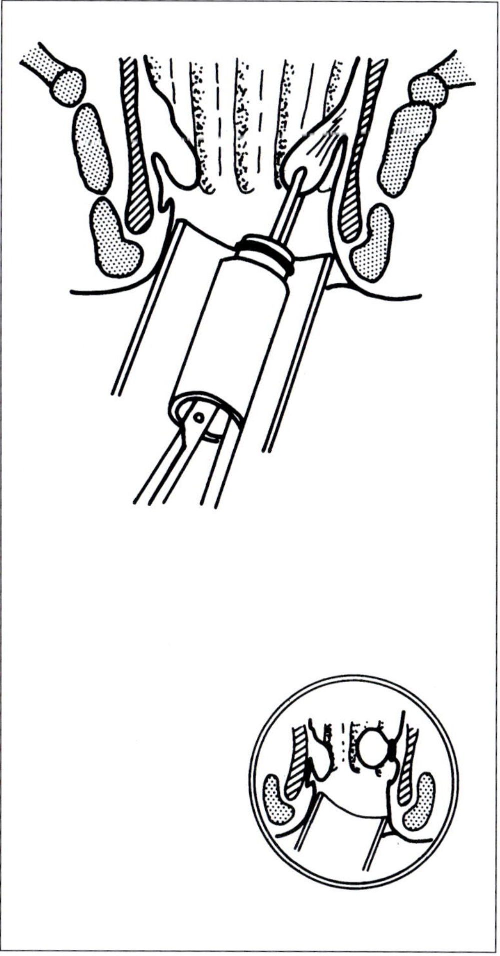

Fig. 5.7. *Use of the McGivney equipment.*

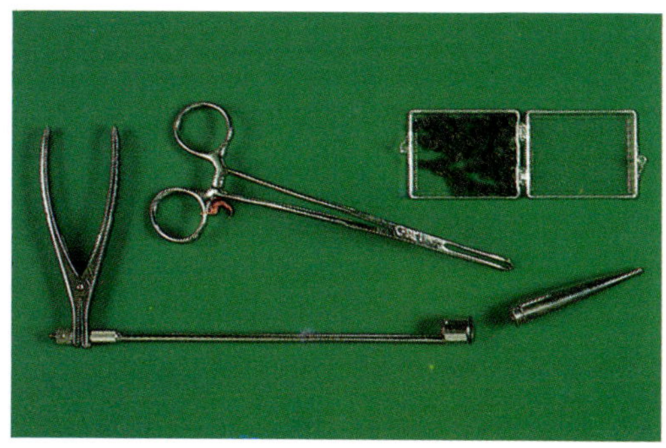

Fig. 5.6. *McGivney equipment for ligation of second and third degree internal hemorrhoids.*

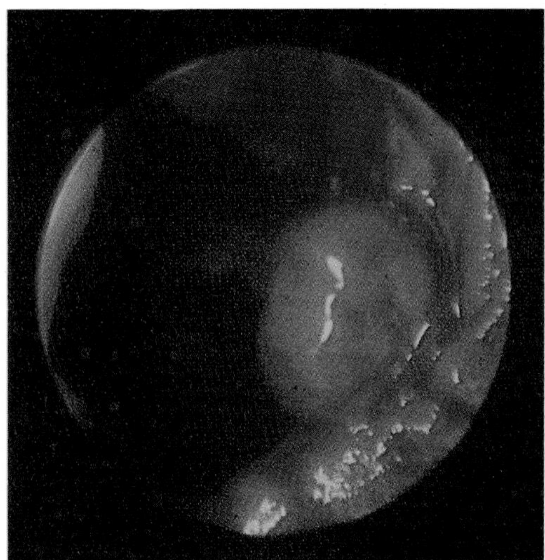

Fig. 5.8. *The immediate appearance of a ligated hemorrhoid.*

large prolapsed hemorrhoids multiple treatment over 2–3 months might be necessary to achieve a satisfactory result. The same contraindications exist as for injection treatment.

Complications. In rare cases a massive secondary bleeding occurs which requires admission to hospital. This is treated by tamponade, electrocoagulation or placement of a new rubber band over the bleeding point. Occasionally edema or thrombosis of external hemorrhoids may also be seen after rubber band ligation. In both cases the patient is best treated with warm sitz baths. If the patient is constipated, prolapse of an internal hemorrhoid with or without thrombosis may occasionally occur. Treatment of this complication is also conservative. Constipation should be avoided in the initial period after rubber band ligation by using a mild stool softener.

Results. Rubber band ligation is a rapid and cheap method for second and in some cases third degree hemorrhoids. Treatment is ambulatory and the patients can return to work immediately after the treatment. About 90–95% of patients with second degree hemorrhoids are cured by this treatment as opposed about 65% of patients with third degree hemorrhoids. The recurrence rate is about 8% a year after rubber band ligation which corresponds to acumulative recurrence rate of 33% after

4 years. The corresponding figures for patients with second degree hemorrhoids which have restored spontaneously are 16% per year and 61% after 4 years. Thus, there is a significantly better prognosis for patients treated by rubber band ligation compared with no treatment at all. The recurrence rate can be significantly reduced by giving the patients long-term dietary fiber supplements after ligation treatment.

There are other forms of conservative treatment for first and second degree hemorrhoids but none of them offer any advantage over injection treatment or rubber band ligation. They will therefore be only briefly mentioned.

Cryosurgery

Cryosurgery is used for first and second degree hemorrhoids. This treatment also aims to prevent prolapse by fixing the tissue to the underlying muscles. The apparatus is relatively expensive and the treatment requires more time than the above-mentioned methods. Furthermore, the treatment results in an unpleasant malodorous discharge from the anus which may persist for many weeks. In our opinion use of cryosurgery for first degree hemorrhoids is an example of overtreatment while its use for second degree hemorrhoids is often unable to prevent continued prolapse.

Infrared photocoagulation

This is a newer form of treatment that may be used for first and second degree hemorrhoids. According to many authors this treatment gives as good results as rubber band ligation. There are practically no complications and no patient discomfort. Coagulation apparatus called Infrarot coagulator MBB-A-T, Munich, is used. The probe is introduced over the hemorrhoid through a rectoscope 1–2 cm above the dentate line where it is pressed against the mucosa. The coagulation time is 1 second and all three hemorrhoids are treated in the same session. The treatment results in coagulation of the tissue with development of an ulcer within about 1 week followed by fibrosis. The disadvantage of this treatment method is the relatively expensive apparatus.

Lord's maximal anal dilatation

This method is recommended by a British surgeon for the treatment of internal hemorrhoids. Its principle is based on the hypothesis that hemorrhoids result from tense fibrous circular bands occurring in both the distal rectum and the anal canal. This stricture formation causes an increased intrarectal pressure with disturbed defecation, increased venous pressure during defecation and thus development of hemorrhoids. The dilatation is carried out under general anesthesia with the patient in the lateral position. Initially two fingers and gradually up to eight fingers are introduced with tearing of the sphincter fibers especially at the 6 and 12 o'clock positions. A finger-sized gauze swab is then introduced into the rectum to prevent hematomas. This swab can be removed after 1 hour. Lord himself recommends in addition that the patient uses a 4 cm thick dilator daily in the first period after the dilatation. Others claim that this is unnecessary. There are many complications following this dilatation treatment: temporary or permanent incontinence for flatus and feces, tears in the perianal skin and mucosal prolapse (especially in patients with third degree hemorrhoids). Because of the many unpleasant complications we do not recommend this method.

Hemorrhoidectomy

Rubber band ligation and injection treatment have reduced significantly the need for formal hemorrhoidectomy in recent years.

There is still, however, an indication for operation with third degree hemorrhoids with large external components, fourth degree hemorrhoids and prolapsed incarcerated hemorrhoids. It is also indicated for patients with accompanying anorectal disorders such as fissures and fistulae, and in cases where other methods, for example rubber band ligation, have been unsuccessful.

Hemorrhoidectomy is painful and requires a long period of sick leave (2–3 weeks) for the patient. It is, moreover, an expensive treatment as many patients require hospital admission for 3–5 days.

The authors prefer and recommend Milligan's method of hemorrhoidectomy which consists of excision of both the external and internal portions of the hemorrhoid and ligature of the pedicel. This is an open method leaving the wound unsutured.

The operation takes places with the patient in the lithotomy position with the buttocks well over the edge of the operating table. General anesthesia is most often used with or without supplementary local anesthesia. Local anesthesia is used if the operation is done as an outpatient procedure. The evening before the operation an enema is given in order to empty the rectum. When the patient has been anesthetized the anal region is washed with soap and water. Anal dilatation is not used in connection with this operation since the advantages of this are small compared to possible complications, such as temporary incontinence or soiling. Kocher's forceps are placed on the skin-covered component of the hemorrhoid in the perianal skin just outside the mucocutaneous junction (Fig. 5.9). Any skin tags should also be included in the forceps. By gently pulling on the forceps the hemorrhoids present themselves into the operation field Artery forceps are now placed on each of the internal hemorrhoids and by traction on these the so-called triangle of exposure is produced. The red rectal mucosa can clearly be seen at the upper pole of the hemorrhoid. The rectal mucosal folds between the three hemorrhoids are also visible. Any small secondary hemorrhoids can be included in the forceps. Having done this the whole operation field is now visible and dissection of the hemorrhoids is not extended any higher up the rectum. A right-handed surgeon starts most easily by dissecting the left lateral hemorrhoid. Both forceps on this hemorrhoid are grasped in the left hand while the index finger of this hand is introduced into the anal canal and pushes the base of the hemorrhoid towards the skin. A V-shaped incision is made in the skin (Fig. 5.9). The subcutaneous perianal space is now exposed and the lowest fibers of the internal anal sphincter can be seen. The dissection continues superficially to the muscle fibers and up towards the tip of the index finger. Thereafter, the mucosa on either side of the hemorrhoid is incised creating a pedicel (Fig. 5.9). The mucosal incisions should radiate in towards the base of the pedicel so that as much mucosa as possible is spared and the pedicle remains as narrow as possible.

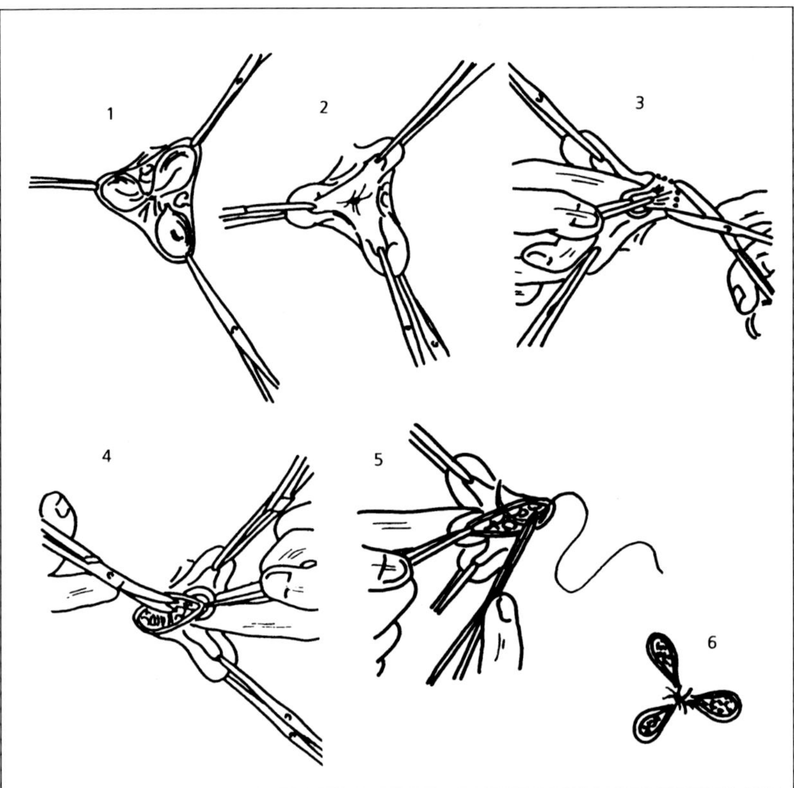

Fig. 5.9. *The principle of Milligan's hemorrhoid operation.*

The pedicel is now transfixed with absorbable 2-0/3-0 sutures. With larger hemorrhoid pedicels a double ligature may be advisable. The hemorrhoid is excised leaving behind a cuff of tissue of about 1 cm distal to the ligature to prevent the ligature from falling off.

The two other hemorrhoids are excised and ligated in the same way. It is very important to preserve intact skin and mucosa bridges of about 1 cm between the hemorrhoids to prevent stenosis on healing. Finally the skin edges are trimmed so that three regular wounds without superfluous skin flaps are left behind. The final result should resemble a clover leaf (Fig. 5.9), where the ligated hemorrhoids pedicels have retracted themselves into the rectum. The area should be dry before dressings are applied. Slight bleeding usually stops after brief compression. Small vessels most often on the skin edges can be electrocoagulated. Gauze swabs are placed on the anal wounds and on top of these a large dressing which is held in place by a T-bandage or net underpants.

Postoperatively the patient is given a mild stool softener. The dressing is left untouched for 24 hours and is thereafter changed twice daily. The patient should have warm sitz baths twice a day. The wounds usually heal within 4–5 weeks. Three weeks after the operation the patient should undergo a rectal examination to check that stricture formation is not beginning.

Complications. Bleeding in the first postoperative day is uncommon and will most often be due to a technical error. Treatment should be compression or, if this is ineffective, electrocoagulation or ligation of the bleeding vessel. Bleeding after about 1 week (1%) is more serious and most often arises from one of the hemorrhoid pedicels. This condition may require blood transfusion if the bleeding is persistent. Such bleeding may be difficult to diagnose initially since blood can accumulate in the rectum and colon. The patient should be fully anesthesized as quickly as possible.

A bivalved speculum is introduced into the rectum, and the blood washed or irrigated away. If the bleeding vessel can be located it should be sutured or ligated with a rubber band. Otherwise, a large Foley catheter with a 30 ml balloon may be introduced into the rectum, the balloon inflated with water and pulled tight. This will usually stop the bleeding. The catheter should remain in place for 48 hours and in a large majority of cases the bleeding will have ceased. Urine retention occurs postoperatively in a few percent of patients. This can nearly always be handled with parasympathomimetic agents. It is very rarely that bladder catheterization is necessary. Stricture formation is a very rare but unpleasant complication after a correctly performed hemorrhoidectomy. The treatment is anal dilation over a long period of time.

Closed hemorrhoidectomy, where the wounds are closed with continuous non-absorbable suture material, is not contributing any advantage over the open excision ligation method. In many cases the suture line disrupts in one or in all the anal wounds. It is a more difficult method and more painful than the open method. Park's submucous hemorrhoidectomy is a difficult and time-consuming operation which does not offer any advantages over the common Milligan operation.

Treatment of prolapsed incarcerated thrombosed hemorrhoids

As mentioned previously this is an extremely uncomfortable and painful condition for the patient. The choice of treatment is between conservative (bed rest, analgesics, warm sitz baths) or acute hemorrhoidectomy. With conservative treatment it takes many days before the pain and edema disappear and more than half the patients will later require hemorrhoidectomy. Therefore, acute hemorrhoidectomy is nearly always employed for incarcerated hemorrhoids. The operation is carried out as previously described but may be difficult because of the edema and the large hemorrhoids. Care should be taken to leave adequate skin and mucosal bridges. It is better to leave behind intermediate hemorrhoids than risking postoperative stricture. Fig. 5.10A–

E shows the principles for the acute operation for a small incarcerated hemorrhoid under local anesthesia.

In order to reduce edema and prolapse injection of hyaluronidase in a solution of 1:200 000 adrenaline/saline can be done just prior to operation. The solution is injected directly into the edematous perianal tissue followed by massage for a few minutes.

The authors use the following treatment criteria for internal hemorrhoids: first degree hemorrhoids can nearly always be treated conservatively without any form of surgical intervention. In the few cases where persistent bleeding occurs sclerotherapy or infrared coagulation may be used. Nearly all second and many third degree hemorrhoids are best treated by rubber band ligation which is easy and cheap and is not especially unpleasant for the patient. Large third degree hemorrhoids and fourth degree hemorrhoids as well as prolapsed incarcerated hemorrhoids are best treated by Milligan's excision ligation method. Only 10% of all hemorrhoids requiring treatment need surgery – the rest can be dealt with by one of the conservative methods.

PERIANAL HEMATOMA
Definition
This condition is also known as thrombosed external hemorrhoids but is in fact a subcutaneous hematoma resulting from rupture of a vein just below the edge of the anal opening.

The hematoma usually arises after a difficult defecation, after lifting a heavy weight or after some other form of vigorous physical work. In some cases no apparent cause can be found for the disorder.

Symptoms and findings
The patient complains of sudden pain in the anal region accompanied by a tender blue–black swelling at the edge of the anus. Pain is exacerbated by defecation and sitting. The hematoma varies from the size of a hazelnut to the size of a grape and is covered by tense skin which may occasionally ulcerate and become necrotic. The diagnosis is usually easy to make but it is important to differentiate a perianal

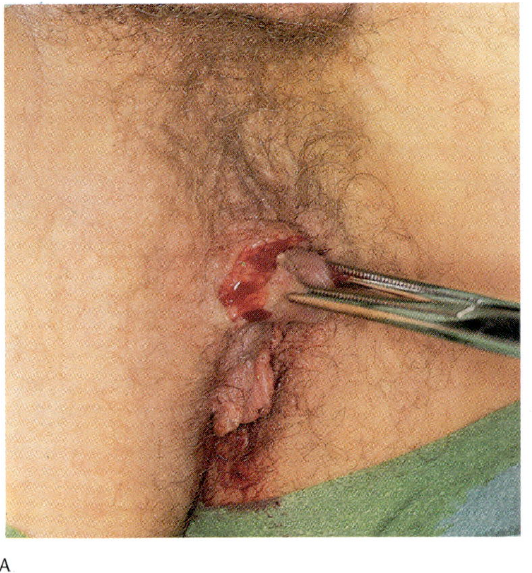

A

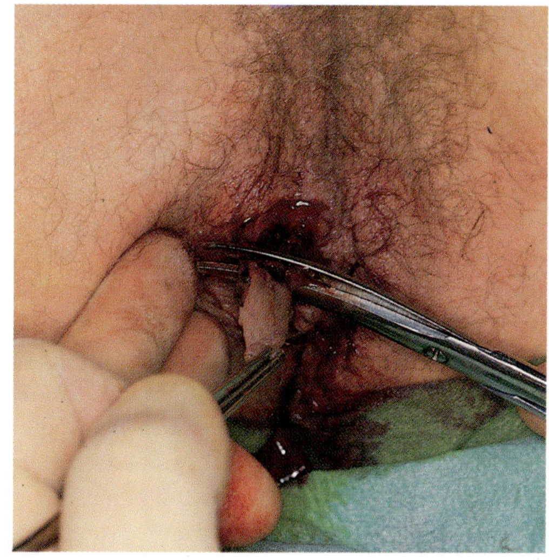

D

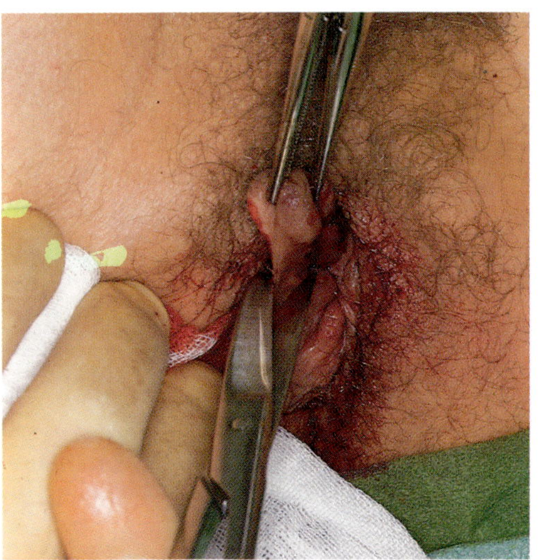

B ⇦   ⇦ C

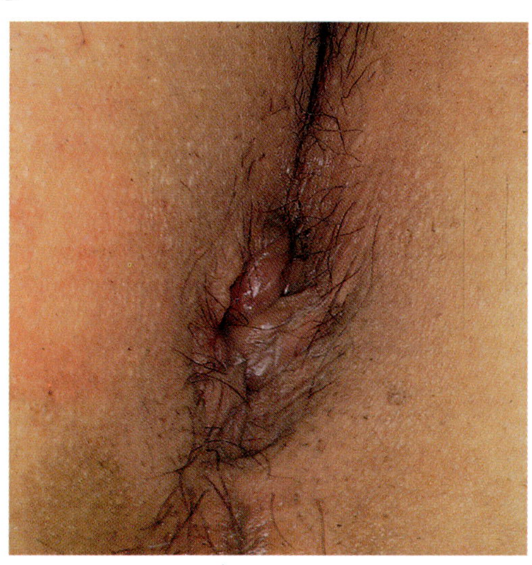

E

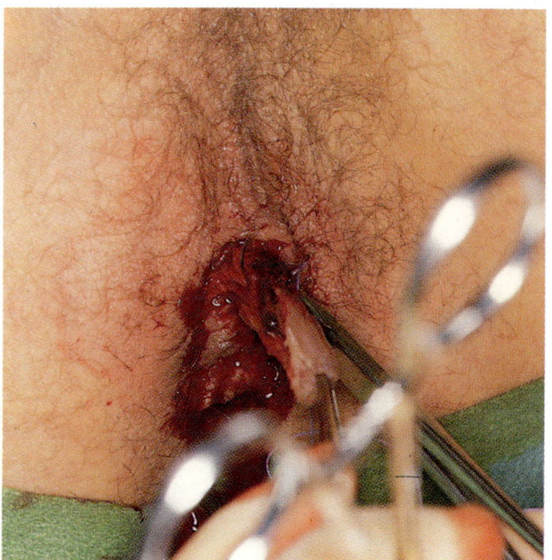

Fig. 5.10. *Surgery for incarcerated hemorrhoid. A. Incision in the skin. B. Separation of layer. C. Dissection of the hemorrhoid. D. Striction and ligation of the top of the hemorrhoid. E. Final result.*

hematoma from an incarcerated prolapsed hemorrhoid since the treatment for the two conditions is different.

Treatment

In the acute stage, within the first 72 hours, when pain is still severe and the swelling is tense, surgery is indicated. The operation consists of evacuation of the hematoma under local anesthesia (Fig. 5.3). The patient is placed in either the lithotomy or the lateral position and local anesthesia is administered with a fine needle around the hematoma. A short (about 1–2 cm) incision is made above the hematoma, whereafter the hematoma can be expressed by squeezing between the thumb and index finger. The wound is left open; a gauze dressing is applied and changed once or twice daily postoperatively. Warm sitz baths are indicated in the first few days after each defecation. The wound usually heals in the course of a week. Postoperative complications are rare. Infection may occur if the hematoma has been incomplete removed. If the hematoma is large the skin edges may end up as small tags which may later require excision if they interfere with wiping after defecation.

If the pain has started to diminish and the swelling has been present for a few days (>72 hours) conservative treatment is indicated. This treatment consists of a laxative and a mild sedative and the pain disappears completely in the course of a few days, and the swelling after a few weeks. After conservative treatment of large perianal hematomas disturbing skin tags may require excision under local anesthesia.

SKIN TAGS
Definition
Skin tags (Fig. 5.11) may occur in the anal region. Idiopathic skin tags are often found without any apparent cause. In other situations they may be the result of a previous perianal hematoma. A special form is the so-called sentinel tag which is found in association with chronic anal fissures. In addition, edematous skin tags and cracks are typical in Crohn's disease and pruritus ani.

Diagnosis
The frequently occurring innocent idiopathic skin tags are usually easy to recognize but it is important to distinguish them from condylomata acuminata, anal cancer (see figures on page 125–126) and the more specialized forms occurring with the chronic fissures, Crohn's disease or pruritus ani.

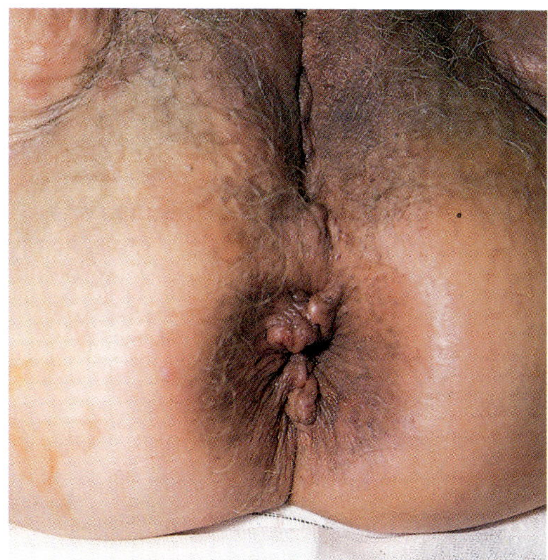

Fig. 5.11. *Anal skin tags.*

Treatment
In the majority of cases no treatment is required. Large skin tags which make cleansing of the anal region difficult should be removed under local anesthesia. The tag should be excised completely leaving a triangular open wound which will heal by granulation. Not uncommonly a large skin tag can occur in association with an internal hemorrhoid and in such cases it is important to undertake an excision ligation of the hemorrhoid which includes the skin tag.

The treatment of skin tags associated with anal fissures and pruritus ani are discussed on pages 52 and 123. Treatment of these skin tags in patients with Crohn's disease or leukemia should be extremely conservative.

NON-SPECIFIC HYPERTROPHIED ANAL PAPILLAE
Definition
This is a an enlarged, often fibrous papilla (fibrous polyp) located at the dentate line.

Anal papillae are seen as small serrations at the proximal end of the anal skin corresponding to the mucocutaneous transition. They give the dentate line its characteristic appearance. Occasionally one of the papillae can hypertrophy and may be 1–1.5 cm long and prolapse into the anus. Such a hypertrophic prolapsing anal papilla often becomes thickened and fibrous and is then called fibrous anal polyp.

Symptoms

Prolapsing hypertrophic anal papillae may give the same symptoms as a prolapsed hemorrhoid. Most often the patient complains of mucus discharge and pruritus ani. Non-prolapsing anal papillae are not thought to give pruritus ani.

Diagnosis

A hypertrophic anal papilla is easily palpable and on anoscopy seen to arise from the tissue of the dentate line.

Treatment

If the hypertrophic anal papilla prolapses and symptoms such as pruritus ani arise, it can easily be removed under local anesthesia. The anal canal is visualized by introducing a bivalvular speculum and the hypertrophic anal papilla is transfixed at its base and removed (Fig. 5.12).

SUPPLEMENTARY READING

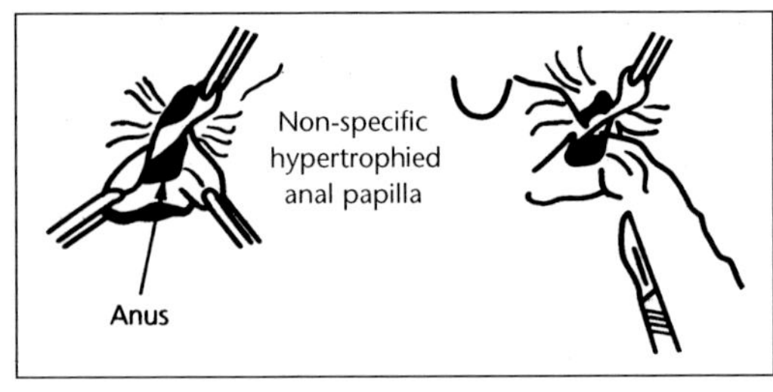

Fig. 5.12. *Non-specific hypertrophied anal papilla and excision of it.*

Ambrose NS, Hares MM, Alexander-Williams J et al. Prospective randomized comparison of photocoagulation and rubber band ligation in treatment of haemorrhoids. Br J Surg 1983; 286: 1389.

Barron J. Office ligation treatment of haemorrhoids. Dis Colon Rectum 1963; 6: 109.

Eisenstat T, Salvati EP, Rubin J. The outpatient management of acute haemorrhoidal disease. Dis Colon Rectum 1979; 22: 315.

Haas PA, Fox TA, Haas GT. The pathogenesis of haemorrhoids. Dis Colon Rectum 1984; 27: 442.

Milligan ETC, Morgan C, Nauton Jones LE, Officer R. Surgical anatomy of the anal canal and the operative treatment of haemorrhoids. Lancet 1937; ii: 1119.

Oh C. One thousand cryohaemorrhoidectomies: an overview. Dis Colon Rectum 1981; 24: 613.

Parks AG. Surgical treatment of haemorrhoids. Br J Surg 1956; 43: 337.

Thomson WHF. The nature of haemorrhoids. Br J Surg 1975; 62: 542.

Wrobleski DE, Corman ML, Veidenheimer MC, Coller JA. Long term evaluation of rubber ring ligation in haemorrhoidal disease. Dis Colon Rectum 1980; 23: 478.

VI. Anal fissures

Definition

An ellipse-shaped defect in the skin-covered part of the anal canal. It is nearly always situated posteriorly (6 o'clock).

Etiology

This is not yet clear but the precipitating factor is thought to be constipation with passage of hard fecal masses. It is however only the case in about 30% of patients. Also patients with diarrhea may develop fissures.

It has not yet been elucidated whether pressure relationships in the rectum play a decisive role in the development of anal fissures. There are as yet rather conflicting results concerning these pressure studies.

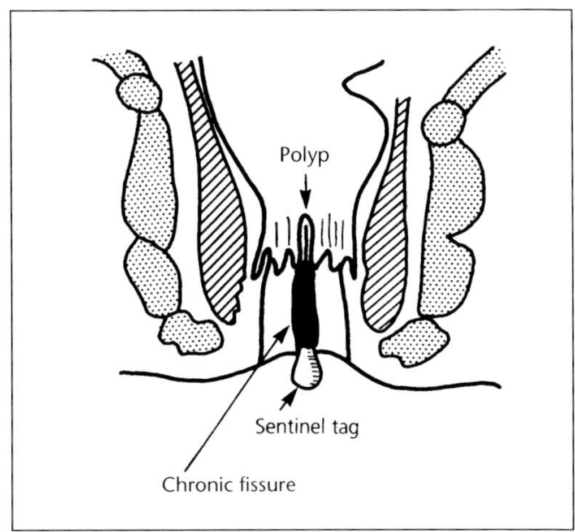

Fig. 6.1. *Schematic drawing of chronic posterior anal fissure with non specific hypertrophic anal papilla (polyp) and a sentinal tag.*

Epidemiology

Chronic anal fissures are thought to occur in men and women in a ratio of 2:1. The median age of the patients is about 40 years. There is a significantly reduced risk of developing anal fissures in patients with high dietary fiber intake compared with those who frequently eat pork products and white bread. Dietary content is thus of significance in the development of anal fissure. There is no correlation between different professions and the development of anal fissures. Consumption of coffee, tea, and alcoholic beverages is also without significance

Pathology

About 98% of fissures in male patients and about 90% of those in female patients are located posteriorly (6 o'clock) in the anal canal. The remaining cases are usually anterior (12 o'clock). Fissures at other sites or multiple fissures occurring simultaneously should arouse the suspicion of a contributory factor such as Crohn's disease. Fissures extend from the anal verge to the dentate line (Fig. 6.1). In the acute form the fissure is a narrow defect in the anoderm (abrasion) located superficial to the internal anal sphincter. In chronic cases the fissure gradually develops into a well-defined wound with fibrotic and sharp edges displaying exposed muscle fibers of the internal anal sphincter. The development of skin thickening (sentinel tag) distal to the fissure because of edema and fibrosis (Fig. 6.1) is also seen with a chronic fissure. A sentinel tag may be the seat of infection and may later develop a superficial fistula (Fig. 6.2). At a late stage in the development of chronic fissures a fibrous polyp might also be seen corresponding to the proximal end of the fissure at the dentate line (Fig. 6.1). This polyp is a hypertrophic anal papilla arising from lymphatic edema and scar tissue.

Characteristic for chronic as well as acute fissures is the accompanying, often extreme, sphincter spasm. It is in fact this severe sphincter spasm towards which the surgical treatment is directed.

Fig. 6.2. *Sentinal tag with fistula formation.*

Symptoms

Anal fissure is the commonest cause of a combination of anal bleeding and severe pain. The main symptom is pain during defecation which persists long after defecation has ended. The pain is described as severe, tearing and burning and the patients tend to avoid defecation and even defer it for several days because of the severe pain. This of course makes the stool even harder and more block-like thus making defecation more difficult and the pain worse. Fresh red bleeding in connection with defecation is a common symptom. A mucus or even a purulent discharge from the fissure can result in soiling and a moist anal region and thus pruritus ani.

Examination and diagnosis

Diagnosis is usually easy. On the other hand, it is very uncomfortable for a patient with a fissure to undergo a regular rectal examination because of the pain. This is also not necessary in the first instance since the diagnosis can be made by gentle retraction of the buttocks after the patient has been placed in the lateral or lithotomy position. This examination would reveal a simple fissure which is always located in the anoderm and as mentioned previously, nearly always posteriorly. The presence of a sentinel tag is a sign of a chronic fissure. If a rectal examination can be carried out without anesthesia, it would reveal severe sphincter spasm. Rectoscopy can usually not be undertaken without anesthesia in the acute stage but this should always be carried out before active treatment of the fissure is instituted.

Differential diagnoses

It is important to reveal whether the fissure is a component of an inflammatory bowel disease, such as Crohn's disease or ulcerative colitis. Malignant tumors in the region can also cause severe pain which resembles fissure pain. Both squamous cell carcinomas and adenocarcinomas of the anal glands may simulate an anal fissure and with the slightest suspicion a biopsy should be taken of the lesion and the inguinal lymph nodes should always be palpated.

Primary syphilis (chancre) may present as a fissure in the anal canal, but is accompanied by bilateral lymphadenopathy. Diagnosis is made by demonstrating spirochaetes in the fissure (see Fig. 9.8). Very occasionally a tuberculous ulcer may simulate an anal fissure.

Chronic pruritus ani with fissure formation in the skin may be misinterpreted as an idiopathic fissure. There is, however, usually more than one rift in the skin and no accompanying sphincter spasm.

Treatment

Conservative treatment

Most acute anal fissures will heal spontaneously with conservative treatment. The aim of treatment is to prevent constipation and diminish pain and muscle spasm with warm sitz baths. Ointments containing local anesthetics or hydrocortisone alone or in combination are often described. These preparations do not work any better than warm sitz baths and fiber supplements. They may, on the other hand, cause contact dermatitis in the anal region and severe pruritus ani. Nevertheless, if ointments are prescribed it is extremely important to instruct the patient in the correct use. It should be pointed out that it is not sufficient to smear the ointment in the anal region but that it should be applied to the actual fissure in the anal canal either with an introducer or a gloved finger. Steroid ointment treatment should not be used for more than 1 or 2 weeks at the most. Warm sitz baths and a simultaneous intake of, 20 g of fiber daily and at least 2 liters of fluid are better than both steroid or anesthetic ointments in the treatment of acute fissures. Long-term fiber treatment of these patients significantly reduces the incidence of recurrence. All patients with anal fisures should be advised to eat about 40 g fiber daily in the form of either fruit or vegetables or a fiber supplement such as Husk, Vi-Siblin, or ordinary wheat bran.

Surgical treatment

Surgical treatment is indicated for chronic fissures and for some cases of acute superficial fissures which do not respond to conservative treatment.

For decades, anal dilatation has been the most commonly used method of treatment for chronic anal fissure. This overstretching of the sphincter paralyses the muscles for many days and leaves an open wound (fissure) which heals from its base in 2 3 weeks. Treatment can be carried out under local anesthesia, but general anesthesia with muscle relaxation is preferred by many. The patient is placed in the lithotomy or lateral position and dilatation is undertaken gradually by first introducing the two index fingers into the anal canal (Figs. 6.3 and 6.4) and then both index fingers and third fingers. A constant traction to the sides is now carried out for 2–3 minutes. The edges of the fissure are thus pulled away from each other and the wound widened. Because of overstretching of the tissue the skin often becomes discolored and in some cases hematomas may develop. There is no postoperative treatment besides daily cleansing with a shower. The method is effective – about 95% of the patients will become pain-free and only about 10–15% will have recurrent symptoms within the subsequent 6 months. On the other hand, the method is associated with many postoperative complications, such as slight incontinence for stool and flatus or soiling, which occur in up to 20% of cases.

Lateral subcutaneous internal sphincterotomy. This operation can be carried out as an outpatient procedure under local or general anesthesia. The patient is placed in the lateral or

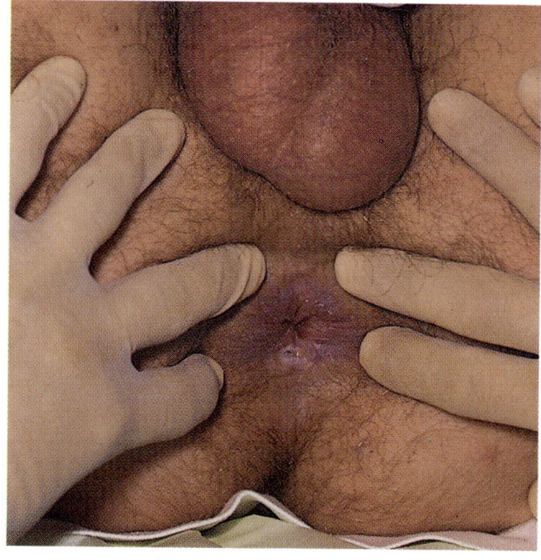

Fig. 6.3. *Inspection of the anal margin disclosing an anal fissure.*

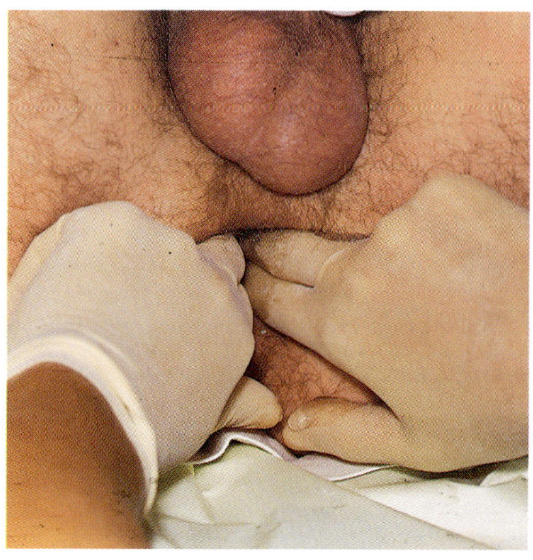

Fig. 6.4. *Moderate anal dialtion with only three fingers (care: never use more than three fingers) with stretching of muscle fibres – especially the interanal sphincter.*

preferably in the lithotomy position. No preparation is required but the operation should be preceded by a rectoscopy. A bivalved speculum is placed into the anal canal and is opened in such a way that the left side of the anal canal is free (Fig. 6.5). The internal anal sphincter can be palpated or even seen. The dentate line can also be identified (Fig. 6.5). A Beaver cataract knife (52L) is introduced through the perianal skin corresponding to 3 o'clock just lateral to the lower end of the internal anal sphincter, in the cleft between the internal and external anal sphincters. A scalpel is introduced between the two muscles up to the level of the dentate line (Fig. 6.6). The scalpel is then passed using small cutting movements through the distal half of the muscle. The innermost muscle fibers lying just under the anoderm are not cut through in order to avoid lesion of the anoderm. After retraction of the knife an index finger is pressed against the mucosa in order to rupture remaining muscle fibers.

Another method is to introduce the Beaver cataract blade medially to the lowest edge of the internal sphincter and pass the blade up between the anoderm and the internal sphincter up to the level of the dentate line (Fig. 6.7). The blade is now turned 90° and with its cutting edge against the muscle fibers the lower half of the internal anal sphincter is cut through by passing the knife laterally through the muscle fibers (Fig. 6.8). A third method is the open technique where a small incision is made in the perianal skin just beneath the lower end of the internal anal sphincter (Fig. 6.9). Using scissors the anoderm is dissected free from the internal anal sphincter. The muscle fibers are white and easily recognizable. Forceps are placed on the lower end of the internal anal sphincter which is also easily dissected free from the external anal sphincter using fine scissors. After isolating the lower end of the internal anal sphincter the muscle is separated with scissors or a knife. 1.5–2 cm of the muscle is transsected. Hemostasis is achieved as a rule by brief compression. Occasionally electrocoagulation or ligation of bleeding vessels is required. The wound is left open without suturing. Any large sentinel tag is excised at the same time as the sphincterotomy. The whole tag is removed but as much of the perianal skin as possible should be spared as should any bared internal sphincter fiber. A large fibrotic hypertrophic anal papilla present at the proximal end of the fissure may be similarly excised. In both cases, however, the fissure itself is left untouched.

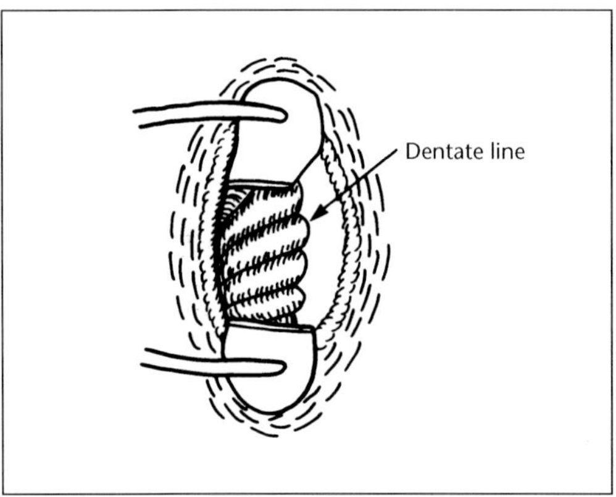

Fig. 6.5. *Exposure of the left part of the anal canal so the dentate line is clearly visible. The patient is ready for a lateral internal subcutaneous sphincterotomy.*

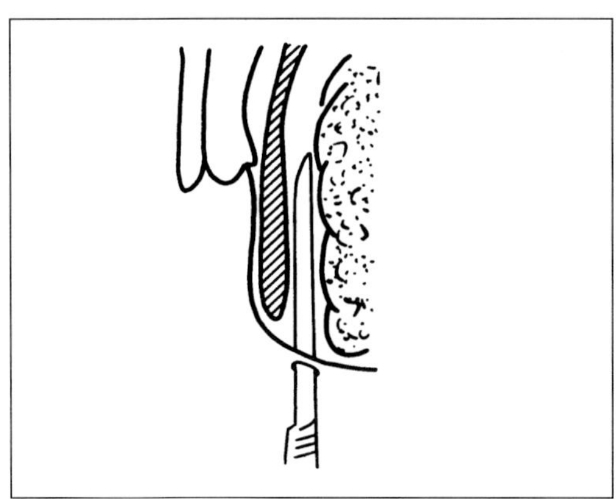

Fig. 6.6. *The performance of subcutaneous lateral internal sphincterotomy with Beavers cateract-knife introduced in the intersphincteric groove until the level of the dentate line.*

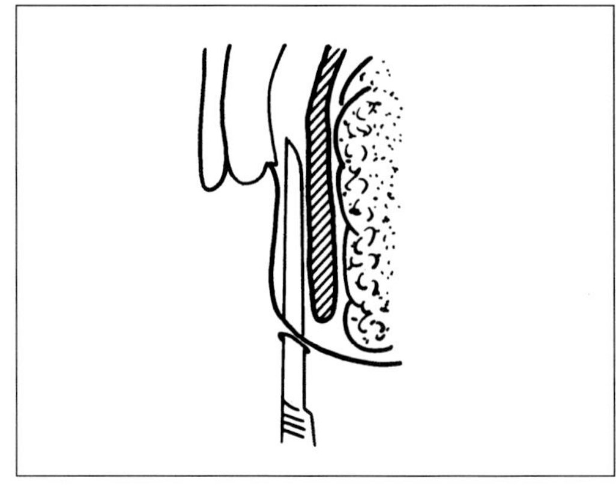

Fig. 6.7. *Subcutaneous lateral internal sphincterotomy performed with the knife introduced in between the anal mucosa of the internal sphincter so that the latter is incised in the direction away from the mucosa – in contrast to the technique shown in Fig. 6.6.*

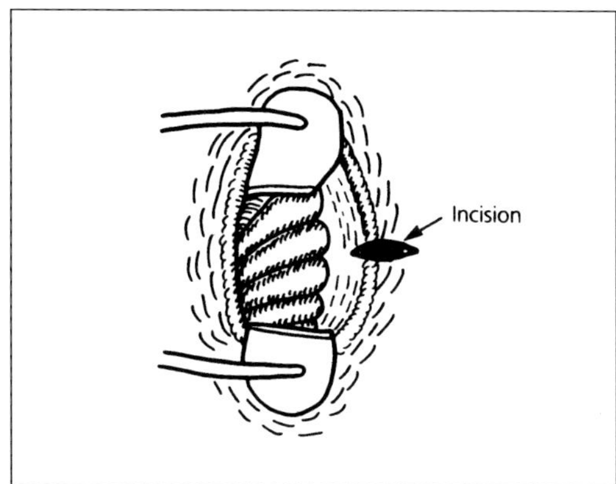

Fig. 6.8. *Open technique: Incision made in the perianal skin and the anal margin so the internal sphincter muscle and the intersphincteric groove can be seen.*

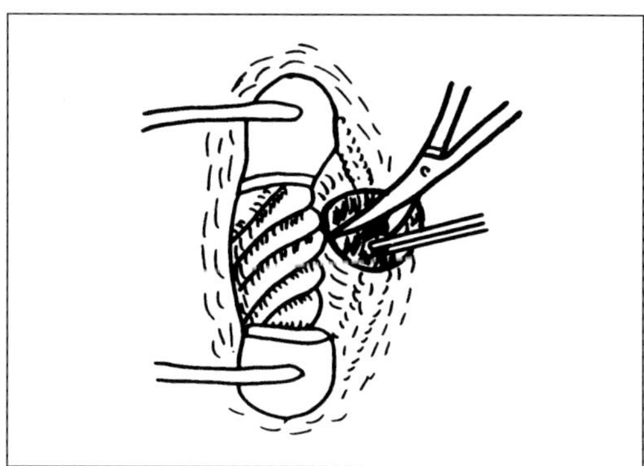

Fig. 6.9. *The distal part (to the left of the dentate line) of the internal sphincter can be cut with scissors as shown in this figure or cut with a knife.*

There is no postoperative treatment besides mild stool softeners for 4–5 days and daily cleansing with a shower.

Complications are rare but bleeding and abscess or fistula formation are described. Furthermore, thrombosis or prolapse of internal hemorrhoids might ensue. It is therefore contraindicated to perform a sphincterotomy (or a dilatation) in the presence of large internal hemorrhoids. In this case the patient should undergo a formal hemorrhoidectomy and sphincterotomy at the same time.

Results after sphincterotomy are good. Over 95% are immediately pain-free and only 5% recur within 6–12 months. Slight incontinence for stool and flatus or soiling occurs less often than after anal dilatation.

Open posterior internal sphincterotomy or triangular excision of the fissure appear to have no place in the treatment of anal fissures because of the many unpleasant postoperative complications.

Prophylaxis

In order to prevent recurrence after both conservative ansd surgical treatment of the fissures it is advantageous to place the patient on long-term fiber supplements. In addition, one can accept that the low plant fiber content of food is a contributing factor to the development of anal fissures and increased intake of plant fiber to about 40 g per day will decrease the incidence of anal fissures in the population.

SUPPLEMENTARY READING

Abcarian H. Surgical correction of chronic anal fissure. Dis Colon Rectum 1980; 23: 31.

Gordon PH, Vasilevsky CA. Lateral internal sphincterotomy: rationale, technique and anaesthesia. Can J Surg 1985; 28: 228.

Jensen SL, Lund F, Nielsen OV, Tange G. Lateral subcutaneous sphincterotomy versus anal dilatation in the treatment of fissure-in-ano in outpatients: a prospective randomized study. Br Med J 1984; 289: 528.

Jensen SL. Treatment of first episodes of acute anal fissure: prospective randomized study of Lignocaine ointment versus hydrocortisone ointment or warm sitz baths plus bran. Br Med J 1986; 292: 1167.

Lewis TH, Corman ML, Prager ED, Robertson WG. Long-term results of open and closed sphincterotomy for anal fissure. Dis Colon Rectum 1988: 31: 368.

Notaras MJ. The treatment of anal fissure by lateral subcutaneous internal sphincterotomy – a technique and results. Br J Surg 1971; 58: 96.

Ray JE, Penfold JCB, Gathright JB, Roberson SH. Lateral subcutaneous internal and sphincterotomy for anal fissure. Dis Colon Rectum 1974; 17: 139.

Rudd WW. Lateral subcutaneous internal sphincterotomy for chronic anal fissure, an outpatient procedure. Dis Colon Rectum 1975; 18: 319.

Watts JMcK, Bennett RC, Goligher JC. Stretching of anal sphincters in treatment of fissure-in-ano. Br J Surg 1964; 2: 342.

VII. Perianal abscesses and fistulae

PERIANAL ABSCESSES
Definition
A collection of pus in a cavity not previously existing in the anal region, usually arising from infection of the anal glands.

PATHOGENESIS AND CLASSIFICATION
It is likely that most perianal abscesses and subsequent chronic anal fistulae arise in the anal crypts. As previously discussed, these anal ducts run from the crypts in the anal canal through the internal anal sphincter and end in the intersphincteric space. Here they form small glandular structures surrounded by lymphoid tissue. The infection begins in an anal crypt and spreads via anal ducts to the intersphincteric space where an abscess forms in the gland (intersphincteric abscess). The infection can spread in various directions, vertically, horizontally, or circumferentially (Fig. 7.1). Perianal abscesses may be classified according to their localization in the perianal space (see Fig. 1.11).

Perianal abscess
The common form, the perianal abscess, arises from vertical spread in a distal direction (Fig. 7.2).

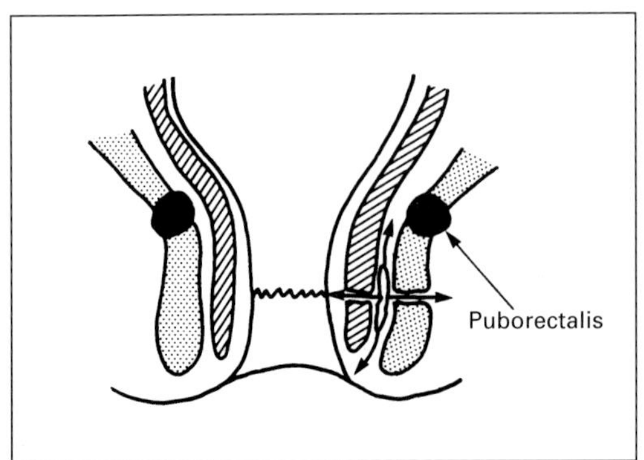

Fig. 7.1. *Anal gland abscesses (intersphincteric abscess).*

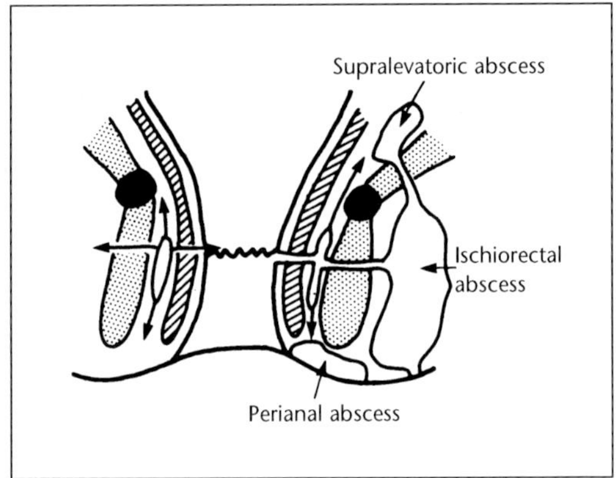

Fig. 7.2. *Possible direction of spread of intersphincteric abscess.*

High intersphincteric abscess – supralevator abscess
This rare high intersphincteric abscess arises by proximal spread. It may be localized either in the rectal wall (high intersphincteric abscess; Fig. 7.3) or located outside the rectal wall and above the levator ani muscle (supralevator abscess; Fig. 7.3).

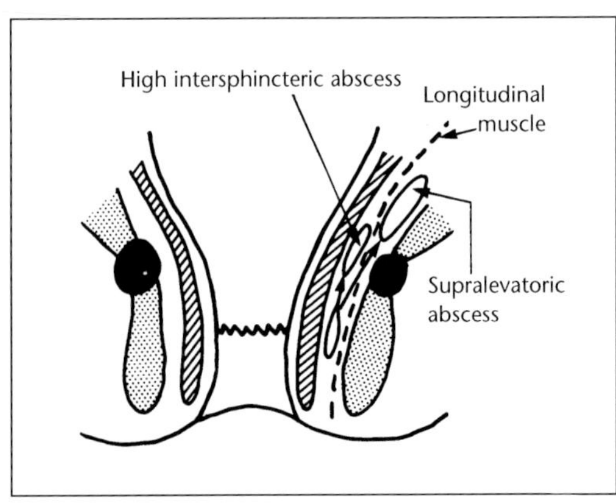

Fig. 7.3. *High intersphincteric abscess (in the rectal wall) and supralevatoric abscess (outside the rectal wall). Both originated from primary intersphincteric abscesses.*

Ischiorectal abscess

The transverse spread of infection can occur in the direction of the anal canal, most often along the course of an anal duct, so that the internal opening of a subsequent fistula may be found at the crypt in the region of the dentate line. It spreads laterally through the anal valves but may also break through the muscle at a higher level (Fig. 7.2). In all cases an ischiorectal abscess arises. Infection can spread from the ischiorectal fossa in a vertical direction downwards with spontaneous rupture through the skin or upwards through the levator ani creating a supralevator abscess (Fig. 7.2).

Horseshoe abscess

The infection can spread in a horizontal direction at different levels, namely, the intersphincteric space, from one ischiorectal fossa to the other or to the supralevator space. These abscesses are described as horseshoe abscesses. The most common form of horseshoe spread occurs between the two ischiorectal fossae and communicates posteriorly via the deep postanal space of Courtney distal to the levator muscles and superior to anococcygeal ligament (Fig. 7.4).

Epidemiology

Anal abscesses are seen more commonly in men than in women (2:1) but the true incidence is not known.

Symptoms

In perianal and ischiorectal abscesses (which are by far the most common forms), acute pain is usually the main symptom. Pain is exaggerated in the sitting position and during defecation. The feeling of a mass in the rectum occurs especially with the perianal localization. If the abscess ruptures spontaneously pus secretion will be noticed together with significant improvement of pain. In large ischiorectal abscesses the patients are usually febrile. The symptoms associated with high intersphincteric abscess are non-specific with the feeling of pressure in the rectum. Such an abscess might spread and present either as a perianal or an ischiorectal abscess.

With a supralevator abscess there is pain both in the perianal and gluteal regions and the patient presents with fever and probably a poor general condition. The diagnosis is difficult and, therefore, the abscess has often already spread to the ischiorectal fossa before the diagnosis has been made.

Diagnosis

A perianal abscess presents with a warm, tender, red and tense swelling close to the anal opening. After a while, obvious fluctuation and possibly skin necrosis with spontaneous rupture occurs. Nothing can be felt on rectal examination.

With an ischiorectal abscess there is no obvious swelling in the anal region but it will instead most often present with a tender swelling of the whole region lateral to the anus. On rectal examination a swelling in the lumen can often be felt. With a horseshoe abscess that has spread to both ischiorectal fossae changes will be found at both sides of the anus and usually posteriorly as well.

The intersphincteric abscess gives no external signs of infection before it has spread and presented as either a perianal or an ischiorectal abscess. On rectal examination a tender swelling in the wall can be felt. This kind of abscess rarely represents more than a few percent of perianal abscesses. Because they are difficult to diagnose and there are few typical symptoms it is only practitioners with extensive experience and specialist interest in these disorders who can diagnose these abscesses in up to 20% of cases.

The supralevator abscess is also difficult to diagnose. If it originates from an intrapelvic condition there will be no external signs of inflammation. Diagnosis is made on rectal or gynecological examination where one can

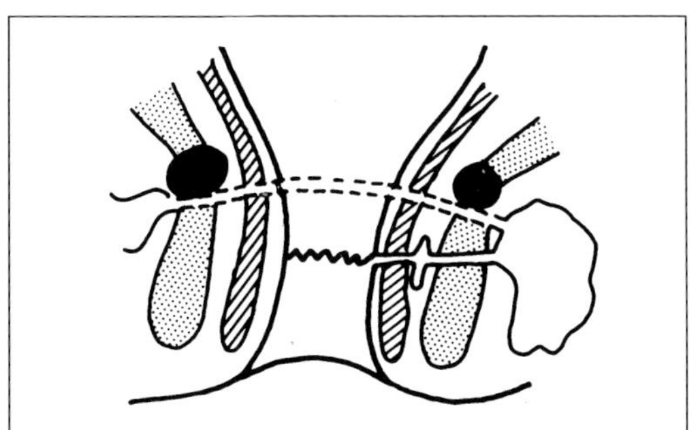

Fig. 7.4. *Ischiorectal horseshoe abscess spreading from one ishiorectal fossa to the other through Courtney's room.*

palpate a tender mass high up in the pelvis. Most supralevator abscesses arise, however, by spreading from either an intersphincteric or an ischiorectal abscess. In the latter case, the diagnosis is made during operation of the ischiorectal abscess when an opening in the levator muscle is found. Transrectal or perianal ultrasound scanning may be useful in the diagnosis of these difficult abscesses.

Differential diagnoses

Hidradenitis suppurativa, infected pilonidal cysts which spread to the anal region, anal fissures with infected sentinel tags and possible fistulation. Also epidermoid and desmoid cysts.

Treatment

Anal abscesses are treated by surgery. The operation can be done under general anesthesia but by far the majority of abscesses, namely perianal and ischiorectal, can be treated easily in an outpatient procedure under local anesthesia. The operation is undertaken with the patient in the lithotomy position with the anal region projecting over the edge of the operating table.

In principle there are two methods of operation: 1. incision and drainage; 2. incision, curettage, and primary suture under antibiotic cover. This holds true for the treatment of the large majority of abscesses, i.e., the perianal and ischiorectal types.

Incision and drainage

For details of local anesthesia the reader is referred to chapter 20. The abscess is opened with a linear incision of about 4 cm. Probes are taken for culture (Fig. 7.5). A finger is introduced into the abscess cavity and all septi are broken down. The skin edges are then trimmed by excising just enough skin to prevent the skin edges from adhering to each other (Fig. 7.5). It is not necessary to perform a large skin excision as this will only delay wound healing. In the acute phase one should not make an intensive search for fistula tracks since this will cause more harm than good by accidentally creating false passages. If there is an obvious fistula opening, for example, a fistula coursing subcutaneously, a fistulectomy can be performed in the acute stage. In other situations a thick silk

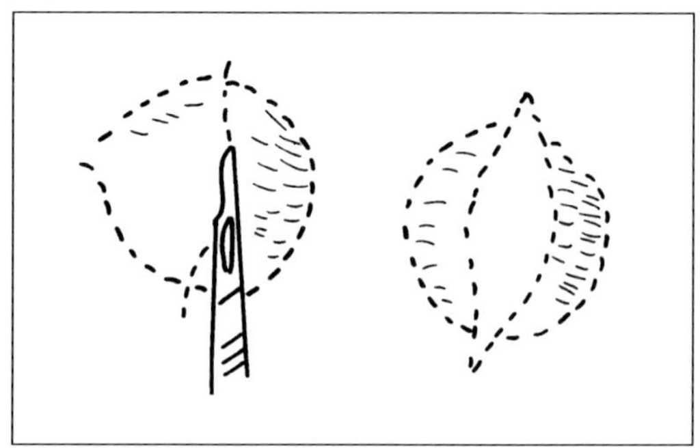

Fig. 7.5. *Incision of an anal abscess with 4 cm long incision and excision of skin edges.*

suture is introduced to mark the fistula tract. When the infection has subsided, after a few days, an experienced proctologist or surgeon can decide what type of fistula is present and, in particular, its relationship to the anorectal ring. It should be noted that in the majority of cases no fistula is found during the primary treatment of the acute abscess. The cavity is loosely filled with gauze which functions merely as a wick. If the cavity is stuffed with dressings this will delay wound healing. Postoperatively the dressings are changed once or twice daily or the cavity is left open and merely covered with a gauze dressing, and the patient may irrigate the wound a few times daily with a shower. Complete healing of the wound usually takes about 4–6 weeks. As mentioned before, in most cases no fistula track is found during the primary acute treatment of the abscess. If the above-mentioned classic open treatment has been used the patient can be examined for fistula tracks a few days later, when the infection has subsided.

Incision, curettage, and primary suture under antibiotic cover

Over the past 20–30 years it has been shown many times that it is possible to treat abscesses, including anal abscesses, by primary suture. This can usually be done as an outpatient procedure under local anesthesia. If a fistula track is discovered the patient should be regarded as unsuitable for primary suture. The abscess is incised and explored as described above. The cavity is then thoroughly curetted to remove granulation tissue, opening blood vessels into the cavity and thereby enhancing access for the prescribed antibiotic. The wound is sutured with 2–4 deep vertical mattress sutures of 2–0 nylon and is covered with a dry, sterile dressing. The sutures are removed after 6–10 days. It is controversial whether one should use the open or the closed method for the treatment of perianal and ischiorectal abscesses. There are more or less the same primary and late results, namely, a satisfactory outcome in 75–80% of the cases after an observation period of 1–2 years. On the other hand, there is no doubt that the primary suture method benefits the patient by greatly shortening the period of treatment. We, therefore, use primary suture under antibiotic cover in most cases, excluding those where a fistula is found at the acute stage.

Concerning the above-mentioned theory about the cryptoglandular origin of anal abscesses the following should be mentioned: if this theory, which appears to be extremely logical, is valid for the majority of anal abscesses and chronic fistulae, it is difficult to explain why most perianal and ischiorectal abscesses can be healed by either the open technique or by primary closure since neither method apparently removes the causative factor of the infection, namely the anal gland.

The treatment of a localized intersphincteric abscess is quite rare. It has often already burst into the rectum making the diagnosis possible when pus is discharged from the rectum. In some cases anal examination reveals such a large perforation in the mucosa that drainage is thought to be sufficient. In the remaining cases where the abscess has not yet burst or where the perforation is small further operative drainage is required. This should be undertaken with the patient fully anesthetized and in the lithotomy position. The abscess can be incised after introduction of speculum into the rectum. This should preferably be done with diathermy since bleeding from the mucosa can be rather heavy. Instead of the standard incision, Goligher has suggested an alternative method which may be useful both in intersphincteric abscesses and fistulae.

Before treating a supralevator abscess it is important to elucidate the cause of the abscess. In most cases one is dealing with the upward spread of an ischiorectal abscess through the levator ani muscle. The supralevator part of the

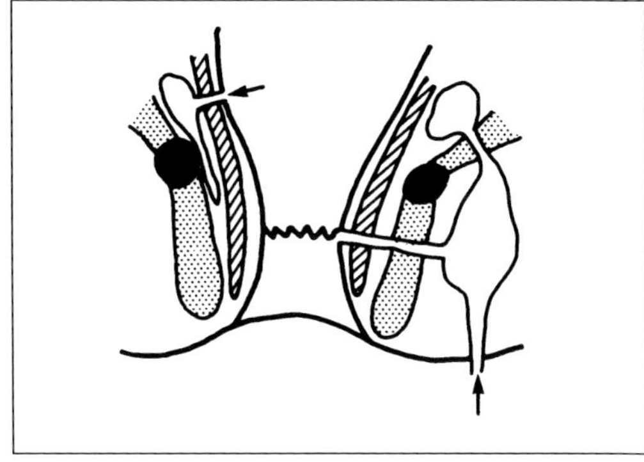

Fig. 7.6. *Drainage of high intersphincteric abscesses into the rectal lumen, and drainage of a supralevatoric abscess through the ischiorectal fossa to the skin.*

abscess is revealed after opening the ischiorectal abscess and finding pus draining from the supralevator space through an opening in the levator muscle. Treatment of a supralevator abscess in such cases is done by widening the defect in the levator and introducing a drain through the muscle into the supralevator space. If such an abscess is drained via the rectum, an extrasphincteric fistula arises which is almost impossible to treat (Fig. 7.6). In the rare case where the suprasphincteric abscess arises by spread of an intersphincteric abscess and an obvious mass can be felt in the rectal wall, the abscess should be drained into the lumen of the rectum. The wall is first punctured with a wide-bore needle and after pus has been collected with a syringe, the abscess is incised and drained via the rectum (Fig. 7.6). After obtaining thorough hemostasis using diathermy a gauze meche moistened in saline is inserted.

ANAL FISTULAE
Definition
An anal fistula is a cylindrical ulcer with an external opening in the perianal skin and an internal opening in the anorectum usually in a crypt at the dentate line. These fistulae can be very complex with multiple branchings.

Etiology
An anal fistula may be regarded as a chronic form of infection in the anal region while abscesses are the acute form. Most fistulae as well as abscesses are of cryptoglandular origin.

It should be remembered that in some cases a specific disease is the origin of an anorectal abscess or fistula. This is true mainly with regard to inflammatory bowel diseases, such as Crohn's disease or ulcerative colitis, but may also hold for abscesses in the pelvis arising from colonic diverticulitis, acute appendicitis, or infections of the internal genitalia.

Rare forms of cancer arising from the anal ducts can present as fistulae (or arise in a chronic fistula).

In the past tuberculosis or actinomycosis were occasional causes for the development of fistulae but these are extremely uncommon causes today.

Classification

It is impossible to understand or treat anal fistulae without adequate knowledge of the anatomy of the anal region (see Chapter 1). It is important to decide the fistula's relationship to the anorectal ring, which marks the transition from the anal canal to the rectum and which is formed by the puborectalis muscle. Fistulae which pass through or over the anal ring are extremely difficult to treat. Transsection of the puborectalis will often cause severe anal incontinence. Transsection of the muscles distal to the anal ring can lead to mild or moderate anal incontinence in certain cases, especially in older patients with weak muscles.

In most cases fistulae have the same origin as abscesses and may thus be regarded as the chronic form of infection in the region. The infection usually starts in the anal glands and spreads from the anal crypts through the internal anal sphincter. In about 80% of anal fistulae the internal opening is found at the dentate line, usually posteriorly. As previously mentioned the infection can spread in three directions vertically upwards or downwards, radially or horizontally (Fig. 7.1).

Intersphincteric fistulae

If the fistula track runs distally from an infection in the anal glands into the intersphincteric space, the commonest form of intersphincteric fistula arises and is thus found distal to the dentate line (Fig. 7.7). The outer opening is found in the perianal skin 0.5–2 cm from the anus. The infection can, however, spread prox-

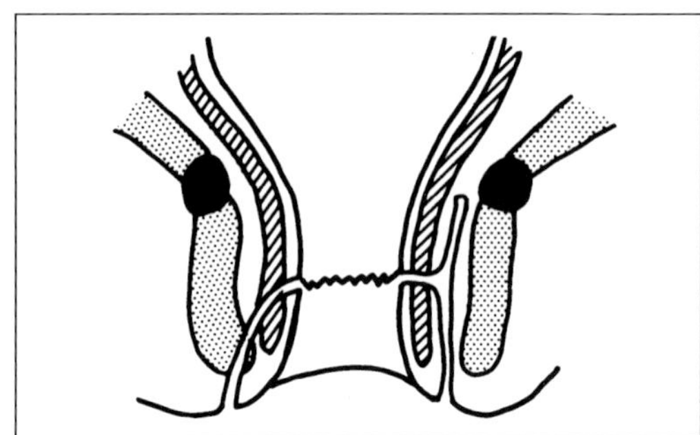

Fig. 7.7. *Simple low intersphincteric fistula and rare variant with a high blind track.*

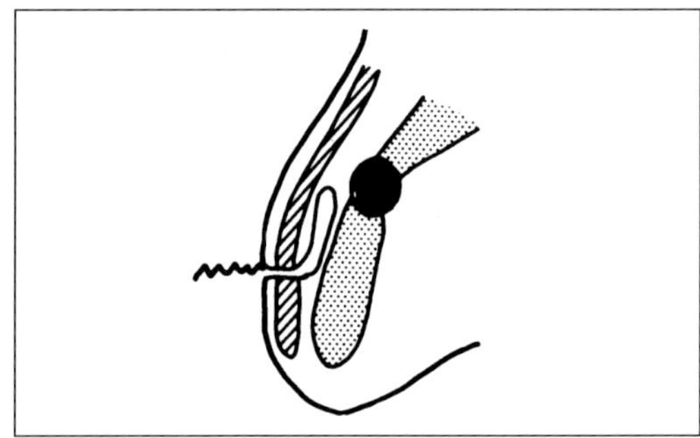

Fig. 7.8. *High intersphincteric fistula without external opening (rare).*

imally in the intersphincteric plane while at the same time having an inferior branch with an opening in the skin (Fig. 7.7). On rare occasions the direction of the fistula is proximal only without an external opening thus constituting a high intersphincteric abscess (Fig. 7.8).

Transsphincteric fistulae

If the infection spreads radially from the primary focus in the intersphincteric space and traverses the external anal sphincter it will reach the ischiorectal fossa (Fig. 7.9). Infection can spread vertically in most cases in an inferior direction with a fistula opening in the skin 2–3 cm from the anus. A transsphincteric fistula with a high blind branch (upward vertical spread) is regarded by Goligher as an ano-rectal fistula as it can often be palpated as an induration in the top of the rectum above the anorectal ring (Fig. 7.9). If this upward extension breaks through the levator ani, a transsphincteric fistula with a high blind

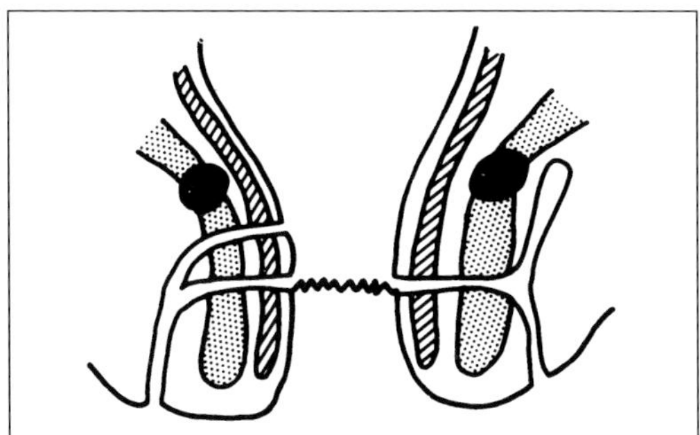

Fig. 7.9. *Transphincteric fistulas.*

supralevator branch arises. Goligher calls this a pelviorectal or supralevator fistula (Fig. 7.10).

A transsphincteric fistula in the ischiorectal fossa often confines itself to one side, but it can spread through the retrosphincteric space of Courtney to the opposite ischiorectal fossa creating a so-called ischiorectal horseshoe fistula

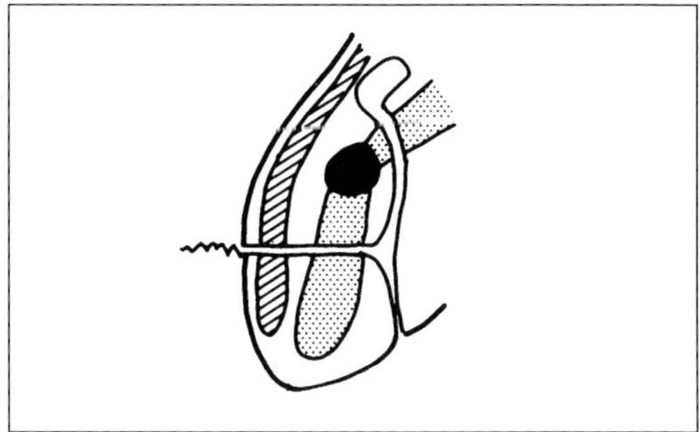

Fig. 7.10. *Transphincteric fistula with high blind supralevatoric track.*

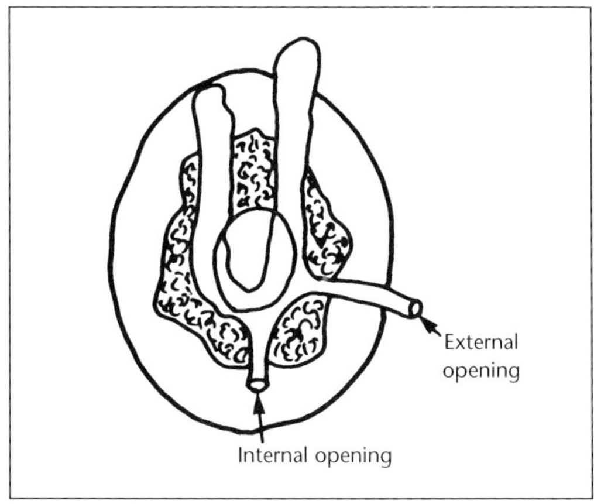

Fig. 7.11. *Ischiorectal horseshoe fistula seen in the horizontal plane.*

(Fig. 7.11). The infection in the two ischiorectal fossae nearly always confines itself distally to the levator ani muscle and the internal fistula opening is nearly always found posteriorly at the level of the dentate line.

All the fistula types named here are transsphincteric in that the primary fistula track passes from the intersphincteric plane through the external anal sphincter. Even though the internal opening in transsphincteric fistula is most often found at the crypts in the dentate line, it should be remembered that fistulae can pass through the external anal sphincter at a higher level but, nevertheless, distal to the anorectal ring (Fig. 7.9).

Suprasphincteric fistulae

These fistulae are extremely rare (2–3%). The fistula begins in the intersphinteric space and passes upwards over the levator ani (Fig. 7.12). From here it courses laterally over the muscle and then downwards between the puborectalis and the levator muscles to the ischiorectal fossa. The infection often spreads in the supralevator space as well (Fig. 7.12).

Extrasphincteric fistulae

These fistulae are also extremely rare (less than 2%) and in fact are not regarded as anal fistulae since they have no communication with the anal canal. These fistulae run directly from the rectum, colon or pelvis through the levator ani and ischiorectal fossa to the skin in the anal region (Fig. 7.13). The fistulae arise most often in connection with diseases such as Crohn's disease, ulcerative colitis, colonic diverticulitis,

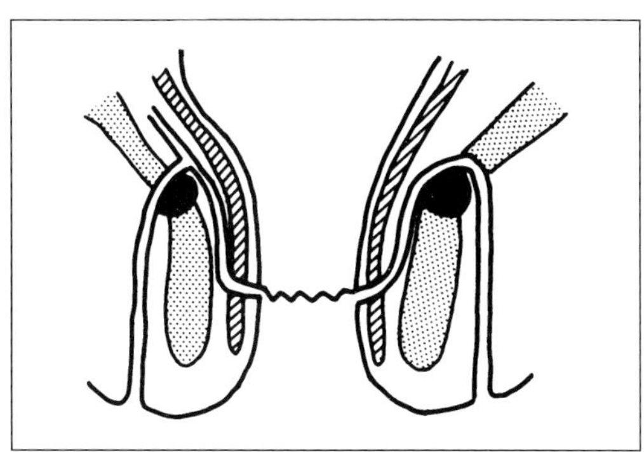

Fig. 7.12. *Supralevatoric fistula. As seen from the figure this fistula spreads superior to the anorectal ring.*

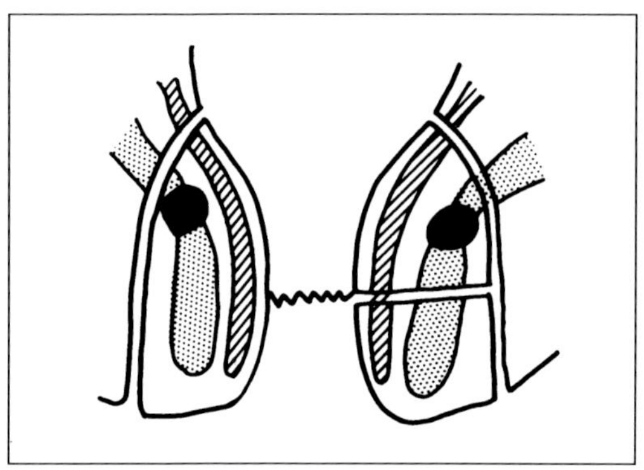

Fig. 7.13. *Extrasphincteric fistula and transphincteric fistula with secondary opening in the rectum (probably always iatrogen).*

or pelvic abscesses. On very rare occasions one may be dealing with a transsphincteric fistula with a secondary opening in the rectum (which is iatrogenic after creating a false passage) (Fig. 7.13).

Symptomatology

Patients have usually had a spontaneous rupture of an abscess or have previously been treated for an anal abscess before the fistula arose. The main symptom of anal fistulae is intermittent or constant discharge of pus, blood, or mucus. Pruritus ani is a common symptom because of the discharge. There is usually no pain except when blockage of the fistula opening results in abscess formation. Those fistulae arising in connection with Crohn's disease may be accompanied by symptoms of the primary disease.

Diagnosis

The patient should be examined in the lithotomy position. Examination should always include: inspection of the anal region, palpation, rectal examination, proctoscopy, and probing of the fistula. The region is examined for external fistula openings. Usually there is only one opening which presents as a small reddish papilla on the skin near the anus. Superficial fistulae can be palpated as an induration in the subcutaneous tissue and one can often express pus from the fistula opening.

On rectal examination the dentate line should be thoroughly palpated for any induration corre-

sponding to the internal opening at the crypts. If the induration is felt higher up it is especially important to determine whether an internal opening exists and its relationship is to the anorectal ring. Anorectoscopy should always be carried out as it might be possible to visualize an internal fistula opening which may discharge some pus during the examination. The examination can be carried out with a bivalved speculum so that one may at the same time carefully evaluate the internal fistula opening from within. Rectoscopy is necessary to reveal diseases of the rectal mucosa, for example, Crohn's disease, ulcerative colitis, or rectal or anal cancer. Subsequent probing of the external fistula opening should be undertaken with the greatest care in order to prevent creating false passages and also to spare the patient any unnecessary discomfort during the examination. Probing of the fistula should in most cases be undertaken under general anesthesia. The distal end of the probe can be bent so that it better suits the estimated location of the fistula. This is especially necessary when examining horseshoe fistulae. During introduction of the probe into the external fistula opening the examiner should palpate the anus with the index finger of the contralateral hand (Fig. 7.14). The point of the probe can, thus, be felt through the rectal wall. In low fistulae the probe soon passes towards the anal canal in an oblique or an almost right-angle direction. In high fistulae the probe passes almost parallel to the rectum for a long distance. It is again impor-

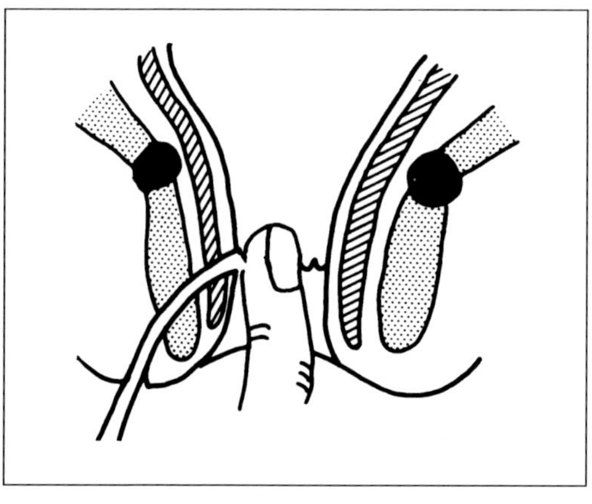

Fig. 7.14. *Probing of fistula with simultaneous palpation of the rectum.*

tant to assess the relationship of the fistula to the anorectal ring by palpating the point of the probe through the rectal wall. It should be stressed that the probe should always be introduced gently and should be passed through the mucosa only if the internal opening has definitely been located.

Fistulography is seldom indicated. This examination can sometimes give further information in the rare cases of high fistulae. X-ray examination of the bowel is indicated if there is suspicion of inflammatory bowel disease. It is as yet unclear whether transanal or perianal ultrasound scanning have any advantage in the investigation of high fistulae.

Treatment

The treatment of anal fistulae is surgery. In many cases it is an exercise which demands great experience and a sound knowledge of the anatomy of the anal region. The operation may result in serious complications, such as severe anal incontinence.

The preoperative preparation consists of an enema, the evening before and another just prior to the operation. The perianal region does not need to be shaved except in very hairy patients. Most fistulae should be operated upon under general anesthesia. The more superfical ones can, however, be handled as outpatient procedures under local anesthesia. Total muscular relaxation should be avoided as it is important to be able to assess the sphincter apparatus's relationship to the fistular track during operation.

The patient is placed in the lithotomy position with the buttocks over the edge of the operating table. The patient is examined again as described previously by inspection, palpation, rectal examination and proctoscopy.

As regards the treatment of anal fistulae associated with specific diseases, such as Crohn's disease, ulcerative colitis, venereal disease and leukemia, the reader is referred to the respective sections on these diseases.

Intersphincteric fistulae

This is the most common fistula and the easiest one to treat. A curved probe is introduced into the external opening and out through the internal opening at the dentate line during simultaneous rectal examination. During the rectal

examination one can palpate the probe through the wall and ensure the fistula is in fact running distally to the anorectal ring. The point of the probe is passed through to the anus (Fig. 7.14). The fistula track is incised as shown in Fig. 7.15 and the probe removed. The open fistula track is curetted in order to remove all granulation tissue. The sore edges are trimmed so that the end result is an open pear-shape wound. After hemostasis a gauze dressing moistened in saline is placed in the wound. The dressing is held in place with a T-bandage or net underpants. During the operation for this type of fistula the distal part of the internal anal sphincter and perhaps the superficial fibers of the external anal sphincter are transsected (Fig. 7.15).

On the other hand, the rare high intersphincteric fistula must be opened into the anal canal (Fig. 7.16). It is important not to confuse such a fistula with the infralevatoric branch of a

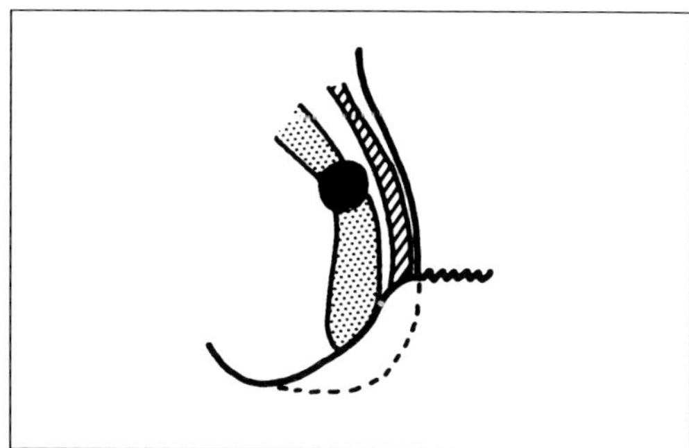

Fig. 7.15. *Illustration of the amount of muscles cut during incision of low intersphincteric fistula.*

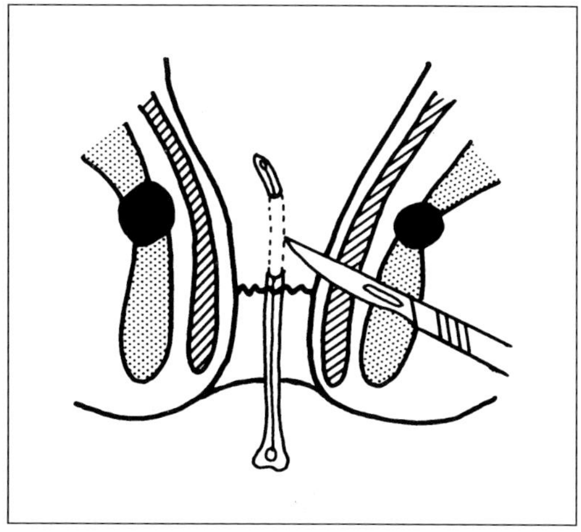

Fig. 7.16. *Probing of high intersphincteric fistula.*

transsphincteric fistula. It would be most unfortunate to open the latter into the rectal lumen. After introducing a bivalved speculum the internal opening at the level of the anal crypts can be probed, and the probe inserted up to the base of the fistula and then through the mucosa into the rectum (Fig. 7.16). Both the mucosa and the internal anal sphincter are incised down to the probe with a diathermy knife. In this way the whole fistula is exposed from the crypt and up through the sphincter muscle. Only the proximal part of the internal anal sphincter is transected in this procedure.

As there is often a fair amount of bleeding from the mucosal edges, Goligher has suggested an alternative method which may also be used for the treatment of a corresponding abscess. Instead of transsection one or two silk ligatures are introduced into the fistula track or the abscess. The ligatures are tied and division of the tissue occurs by necrosis in the course of 4–6 days without significant bleeding.

Transsphincteric fistulae

Most of the fistulae pass through the external anal sphincter at the level of the dentate line and can, therefore, be treated by incising over a probe with transection of the distal ends of the internal and external anal sphincters (Fig. 7.17). If the fistula track passes through the external anal sphincter at a higher level (Fig. 7.9) but still below the puborectalis muscle, it may be necessary to transect the greater part of the external anal sphincter. In most cases an intact puborectalis muscle will be able to maintain continence but, particularly among older patients with weakened musculature, there is a risk of inconti-

nence of varying degrees if most of the external sphincter is transsected. It may, therefore, be necessary to use a Seton suture for the proximal part of the external anal sphincter. In those cases where the transsphincteric fistula has spread upwards to the apex of the ischiorectal fossa or perhaps broken through the levator ani muscle (Fig. 7.10) these fistula tracks must be drained. The hole in the levator ani muscle must be dilated so that the infection in the supralevator space is given an outlet (Fig. 7.18) Caution should be taken when probing these high extensions in order not to perforate the rectal wall. The most difficult thing with the latter type of fistula is to find the primary fistula track which arises at a right-angle from the long vertically coursing fistula track. It is, however, important to identify this fistula track possibly by retrograde probing since neglecting to open this part of the fistula will result in recurrence.

Since the primary ischiorectal fistula track follows the lower edge of the puborectalis muscle spread to the contralateral side will cause a horseshoe fistula pattern. When treating such a horseshoe fistula the probe is carefully introduced into the external fistula opening. The probe is passed almost parallel to the anal canal up to the lower edge of the anorectal ring taking care not to perforate the rectal wall. The anteriorly coursing part of the fistula should be incised first. It is easiest to carry out the incision if the curved probe is forced through the skin at a point anterolater-

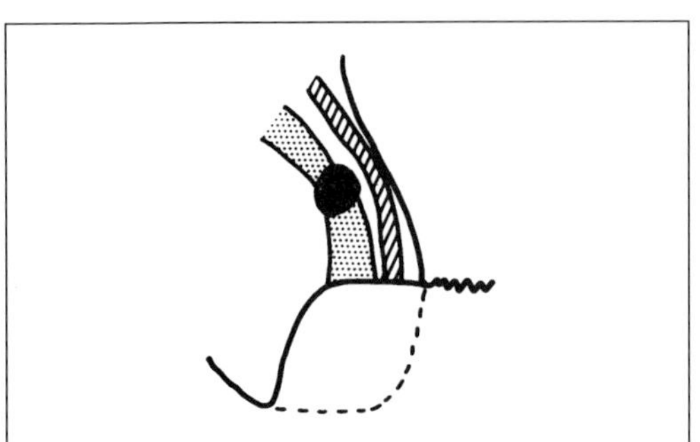

Fig. 7.17. *Illustration of the amount of muscle cut during incision of transsphincteric fistula.*

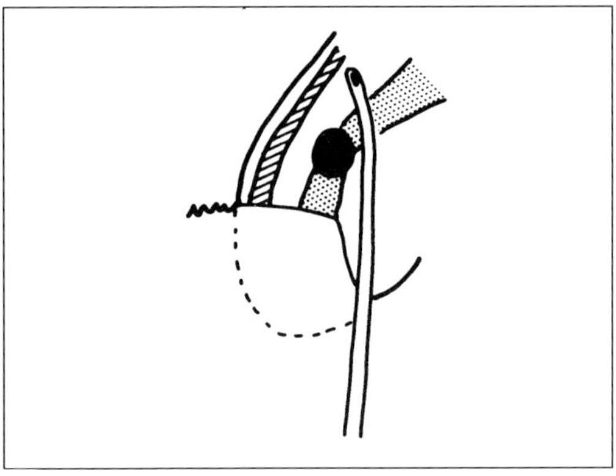

Fig. 7.18. *Treatment of transsphincteric fistula with supralevator track. The transsphincteric part is treated as usual with devision of the lower parts of the internal and external sphincter muscles.*

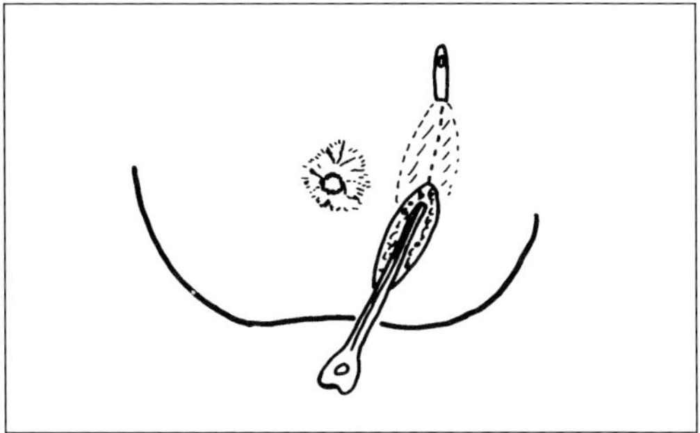

Fig. 7.19. *Treatment of ischiorectal horseshoe fistula. The superior part of the fistula on the same side as the external opening is incised.*

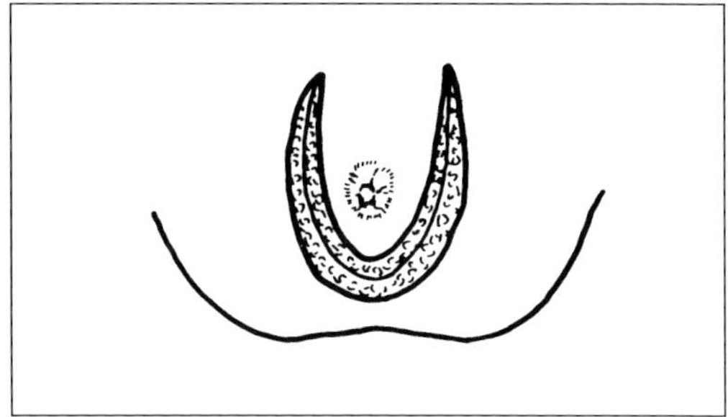

Fig. 7.21. *Treatment of ischiorectal fistula. The anterior part of the fistulous track on the opposite side is incised and a high horseshoe shaped wound is performed after the skin edges have been excised.*

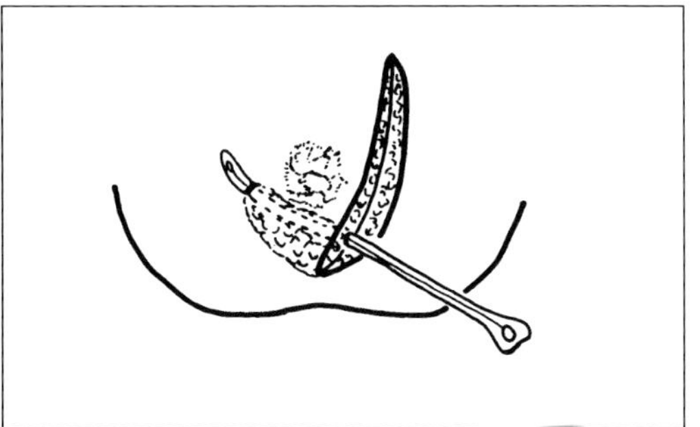

Fig. 7.20. *Treatment of ischiorectal fistula. The postanal part of the fistulous track is incised.*

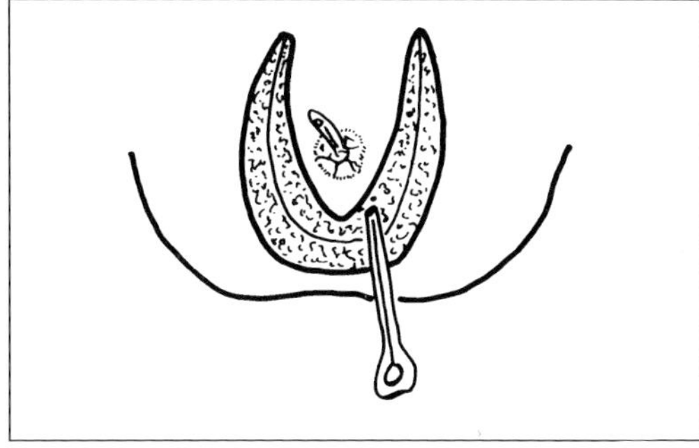

Fig. 7.22. *Treatment of ischiorectal fistula. Probing the fistula track to the anal canal.*

ally to the anus (Fig. 7.19). An incision is then made down to the probe and a section of this branch of the fistula is laid open. After achieving hemostasis and trimming the skin edges the remaining part of the anteriorly running fistula track is incised. The connection running posteriorly to the anus to the fistula track on the contralateral side is now located (Fig. 7.20). Similarly this part of the fistula system is most easily opened after forcing a probe through the skin. Once again hemostasis is achieved and the skin edges trimmed.

The same procedure is carried out on the anteriorly running fistula on the opposite side (Fig. 7.21). In some cases there is a fistula extension running posteriorly towards the coccyx. If such a track is found it is probed and opened. At this stage it is of great importance to find any connection to the anal canal which will usually have an internal opening sited at a crypt posteriorly in the anal canal (Fig. 7.22). This section is now incised by transsecting the involved part of the sphincter musculature. One might instead choose to employ a Seton ligature through the muscles in the first instance (Fig. 7.23). It should be emphasized that the above mentioned operative procedure can be very difficult and that it is always a major undertaking with a fair amount of bleeding from vessels in the subcutaneous fat. It is necessary to take great care during probing of the different fistula tracks to avoid creating false passages. Painstaking hemostasis should be carried out during the procedure, and the wounds should be trimmed successively as shown in the illustrations. All granulation tissue must be removed by curettage.

Suprasphincteric fistulae

In these cases the usual technique of opening the whole fistula track by transsecting the involved muscle cannot be used as this will surely lead to incontinence. The treatment consists of dividing the lower half of the internal and external anal sphincters and placing of a Seton ligature around the distal part of the involved sphincter muscles and the puborectalis (Fig. 7.24). The Seton

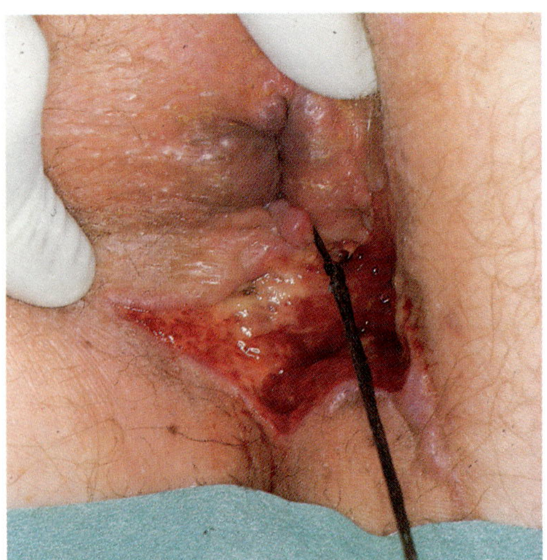

Fig. 7.23. *Treatment of ischiorectal horseshoe fistula. Seton-ligature in place.*

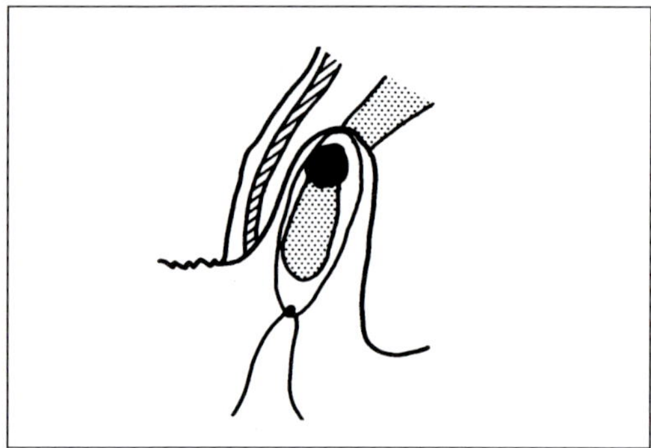

Fig. 7.24. *Treatment of supralevatoric fistula. The lower halves of the internal and external sphincter muscles have been divided and a Seton-ligature is placed around the upper part of the external sphincter and puborectalis.*

suture functions as a drain and after 8–12 weeks when the anal wound has safely healed around the suture, the suture can simply be removed. We do not recommend transsection of the puborectalis muscle even in a gradual fashion as this is accompanied by subsequent incontinence.

Extrasphincteric fistulae

Extrasphincteric fistulae can be treated neither by the open technique nor by the Seton technique. If the fundamental cause is a bowel disease, such as Crohn's disease or colonic diverticulitis, the primary lesion must be treated by, for example, bowel resection or excision of the rectum. If the fistula is secondary to a pelvic abscess that has broken through the levator ani into the perineum, treatment should be directed towards the abscess in the pelvis.

In the remaining very rare cases where a transsphincteric fistula has a secondary internal opening in the rectum, the course is apparently always iatrogenic (Fig. 7.13). The secondary opening in the rectum is a result of exaggerated fistula probing with perforation of the wall. The ischiorectal part of the fistula system is treated by opening the fistula by transsecting the lower half of the internal and external anal sphincters. On the other hand the proximal parts of these muscles are spared in order to maintain continence. The hole in the levator ani is widely dilated so that the opening in the rectal wall can be visualized and sutured with non-absorbable material. In some cases the rectal perforation can be closed via the lumen of the rectum.

Before this operative procedure a defunctioning end sigmoideostomy is established. The wound is trimmed and is allowed to granulate from its base in the usual manner.

If an extrasphincteric fistula gives only minimal symptoms, for example intermittent discharge, one may opt for the expectant approach even for many years.

Use of a Seton ligature

It is difficult to define precisely in all cases how much sphincter musculature one can transsect while treating anal fistulae without compromising continence. Even transsection of only the lower half of the internal anal sphincter plus superficial fibers of the external sphincter may result in mild incontinence in some cases. In the large majority of cases, division of the lower half of both the internal and external sphincters will not result in any form of incontinence. One should, however, be careful with older patients who have weaker anal muscles. It should, moreover, be remembered that the puborectalis muscle is absent anteriorly so that when dealing with an anterior fistula as many fibers of the external sphincter as possible should be preserved. When a fistula track crosses the sphincter at a high level, one can use the so-called Seton ligature, which consists of a heavy silk or nylon suture.

A Seton ligature may be used primarily to mark the fistula track passage through the muscles allowing the surgeon to evaluate more easily the fistula's relationship to the anorectal ring, when the patient is awake.

Fistulae which pass through the proximal part of the external anal sphincter or just over the anorectal ring are suitable for treatment with a Seton ligature. The ligature is inserted around the proximal part of the muscle after the distal part has been transsected (Fig. 7.25). The ligature should not be tightened. It is merely tied loosely. The ligature causes fibrosis of the musculature around the suture so that the muscle does not gape after a later transsection. Furthermore, and possibly most important, this suture functions as a drain for the anal wound which heals slowly around the Seton ligature. In a few cases the ligature saws its way through the muscle. The suture should remain in place for a long period, 10–14 weeks, in which time the anal wound will have healed in the majority of cases and the suture can be removed. In a few cases where this does not happen some recommend transsection of that part of the

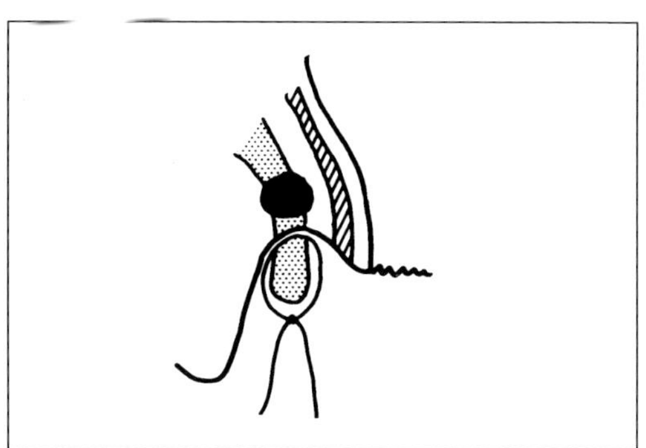

Fig. 7.25. *Placement of a Seton around the upper part of the external sphincter.*

muscle which is comprised by the ligature. This can only be done with a rather high risk of developing a greater or lesser degree of incontinence. Many patients have developed incontinence and we are of the opinion that the puborectalis should not be transsected.

SUPPLEMENTARY READING

Bennett RC. A review of the results of orthodox treatment for anal fistulae. Proc R Soc Med 1962; 55: 756.

Christensen A, Nilas L, Christiansen J. Treatment of transsphincteric anal fistulas by the Seton technique. Dis Colon Rectum 1986; 29: 454.

Eisenhammer S. The final evaluation and classification of the surgical treatment of the primary ano-rectal cryptoglandular intermuscular (intersphincteric) fistulous abscess and fistula. Dis Colon Rectum 1978; 21: 237.

Goligher JC. Surgery of the anus, rectum and colon, 5th ed. London: Balliere Tindall, 1984.

Henley PH, Ray JE, Pennington EE, Grablowky OM. Fistula-in-ano: a ten year follow-up study of horseshoe abscess fistula-in-ano. Dis Colon Rectum 1976; 19: 507.

Held D, Khubchandani I, Sheets J et al. Management of anorectal horseshoe abscess and fistula. Dis Colon Rectum 1986; 29: 793.

Kronborg O. To lay open or excise a fistula-in-ano: a randomized trial. Br J Surg 1985; 72: 970.

Kuypers JHC. Diagnosis and treatment of fistula-in-ano. Neth J Surg 1982; 34: 147.

Marks CG, Ritchie JK. Anal fistulas at St. Mark's Hospital. Br J Surg 1977; 64: 84.

Parks AG, Sitz RW. The treatment of high fistula-in-ano. Dis Colon Rectum 1976; 19: 487.

Park AG, Gordon PH, Hardcastle JD. A classification of fistula-in-ano. Br J Surg 1976; 63:1.

Vasilevsky C, Gordon PH. Results of treatment of fistula-in-ano. Dis Colon Rectum 1985; 28: 225.

VIII. Pilonidal sinuses and cysts

Definition

One or more midline openings in the gluteal cleft that extend into the subcutaneous layer at various depths as epithelial or granulation tissue-lined fistula tracks (pilonidal sinus) where they end blindly or in a cystic cavity (pilonidal cyst). One or more fistula tracks may be found to have secondary skin openings lateral to their primary opening in the gluteal cleft. Symptoms arise when the primary or secondary fistula track becomes blocked and infected, possibly with resulting abscess formation.

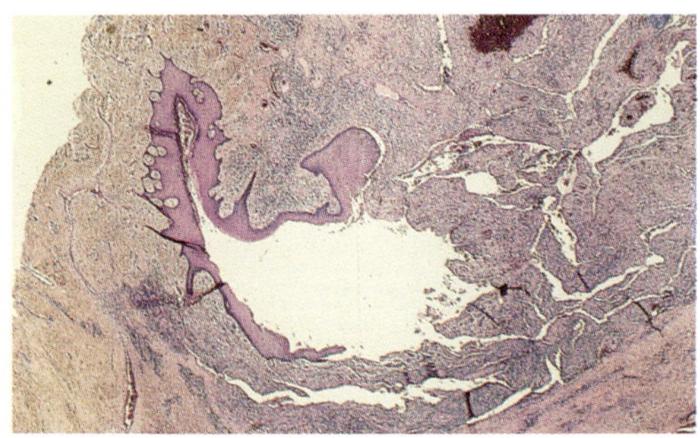

Fig. 8.1. *Histological picture of pilonidal cyst with granulation tissue.*

Epidemiology

The incidence of the condition is not known with certainty. In an American examination of about 32,000 male and 21,000 female students without symptoms and with an average age of 21 years, 364 men and 24 women were found to have pilonidal sinuses. The male:female ratio of patients treated for symptomatic pilonidal sinuses or cysts is given in the literature as between 3:1 to 7:1. The disorder is extremely rare in patients under 17 years of age and occurs most frequently between the ages of 20 and 25 years, after which the incidence again falls rapidly.

Pathological anatomy

Several histological examinations have shown that the primary sinus is often lined with well-defined squamous epithelium both proximally and distally, while if a cyst cavity is present it is lined by granulation tissue (Fig. 8.1). Other investigations have, however, been unable to show an epithelial lining in the primary sinus. The secondary fistula tracks and openings are lined by granulation tissue, which is often seen to extrude from the secondary skin openings laterally to the midline in the gluteal cleft. In about half the patients, hair can be demonstrated in the primary sinus lying loosely in it or attached to the skin surrounding the sinus and growing down into it (Fig. 8.2).

Etiology

There have been many theories of the etiology of this condition since it was first described by Mayo in 1833, but its etiology is still uncertain. In 1850, Hodges introduced the name pilonidal sinus (*pilus* = hair, *nidus* = nest) for the condition which until 1946 was regarded as a congenital disorder.

On the basis of various theories, it has been accepted since 1946 that the condition is not congenital but acquired. The disorder has also been called jeep disease, since it was very common amongst short-haired American soldiers during the Korean War. It was thought that hair from these cropped soldiers was shed between the buttocks and into the small sinus openings. However, since hair can be demonstrated in the sinuses in only half of the patients, the theory that a pilonidal sinus develops from a hair follicle and that the midline openings represent inverted hair follicles is preferred.

The condition is seen most frequently in the gluteal cleft, but has also been described in the axillae, clitoris, umbilicus, sole of the foot, and anal canal as well as between the fingers of barbers. Pilonidal sinuses and cysts containing wool, grass, animal hair, and hair of a color different from the patient's have also been described.

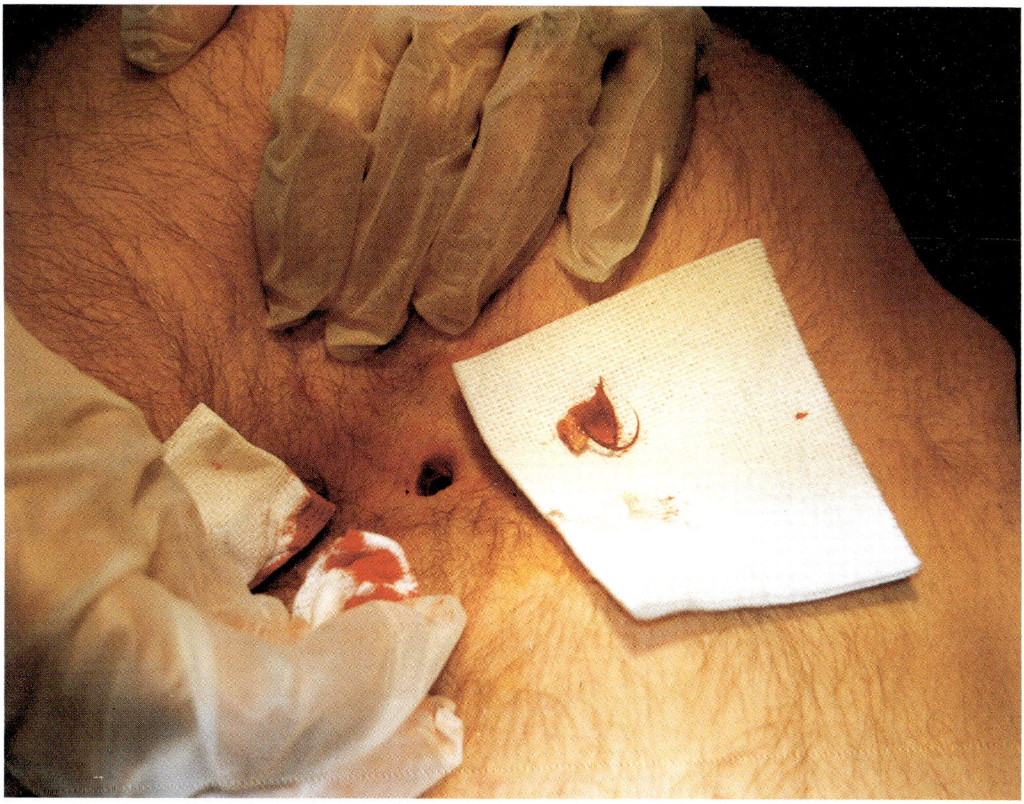

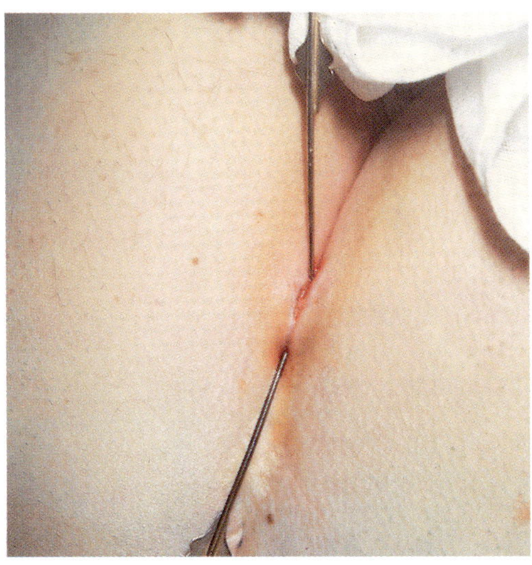

Fig. 8.2. *Excision of primary sinus with hair "pointing" to the cyst-cavity.*

Symptomatology

The first episode of this disease presents most frequently as an acute pilonidal abscess and less frequently as a chronically discharging infected sinus or pilonidal cyst. Acute pilonidal abscesses are most often followed by recurrent abscesses with chronic infection in the sinus or the cyst in the intervening periods.

Acute pilonidal abscess

This disease develops and presents most often in the same way as an acute abscess in other parts of the body, with tenderness, pain, swelling, redness and feeling of warmth. The abscess is often located relatively deeply under thick skin, so that spontaneous rupture rarely occurs. The condition is often accompanied by fever, leukocytosis and general signs of infection.

Chronic pilonidal sinus and cyst

Less frequently the condition presents primarily as a chronic infection with discharge from a fistula track and/or cyst cavity (Fig. 8.3).

The most common presentation is recurrent abscess formation followed by persistent or intermittent discharge from a chronically infected fistula track.

Fig. 8.3. *Chronic pilonidal sinus.*

Malignant transformation

There are cases described of malignant transformation of a chronic pilonidal sinus or cyst to a squamous cell carcinoma although this is extremely rare. We have never seen such a case but recommend, nevertheless, that all tissue removed from a sinus or a cyst is sent for histological examination.

Diagnosis

Diagnosis is usually easily made from the physical signs. One should, however, bear other diagnoses in mind. Similar conditions are hidradenitis suppurativa or a simple carbuncle or furuncle. Extremely rarely one may see an abscess-like lesion in the gluteal cleft, that is caused by osteomyelitis of an underlying bone. One should exclude that the condition is caused by an anal fistula which may extend posteriorly before ending in a retrorectal abscess. If probing of the fistula opening shows that it is coursing away from the anus, this suggests a pilonidal abscess. Pilonidal sinus or abscess may be seen together with an anal fistula.

Treatment

Acute pilonidal abscess

An acute pilonidal abscess (Fig. 8.4) often presents as an extremely painful lesion and rapid incision will give immediate relief (Fig. 8.5).

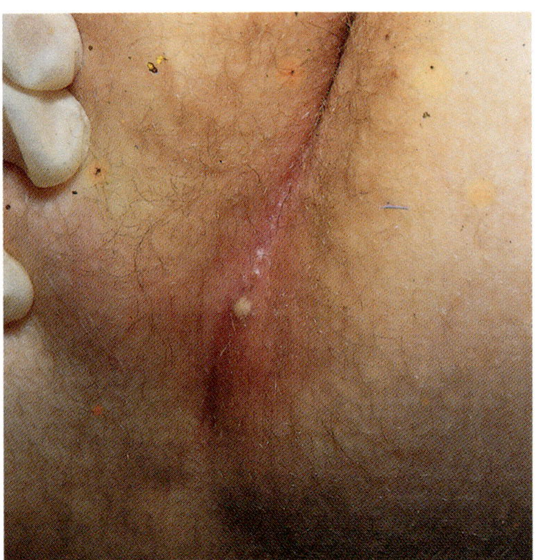

Fig. 8.4. *Acute pilonidal abscess.*

Patients with acute pilonidal abscesses should be treated in the first instance by simple incision and drainage of the abscess. Most can be treated as outpatients under local anesthesia injected well away from the abscess. A cruciate incision is made (Fig. 8.6), and drainage is instituted by breaking down any septa with a curette. The corners of the incision are excised

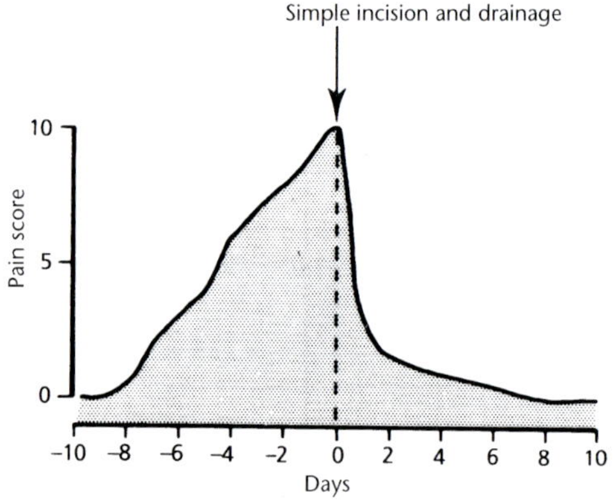

Fig. 8.5. *The effect of emergency incision of pilonidal abscess on pain relief.*

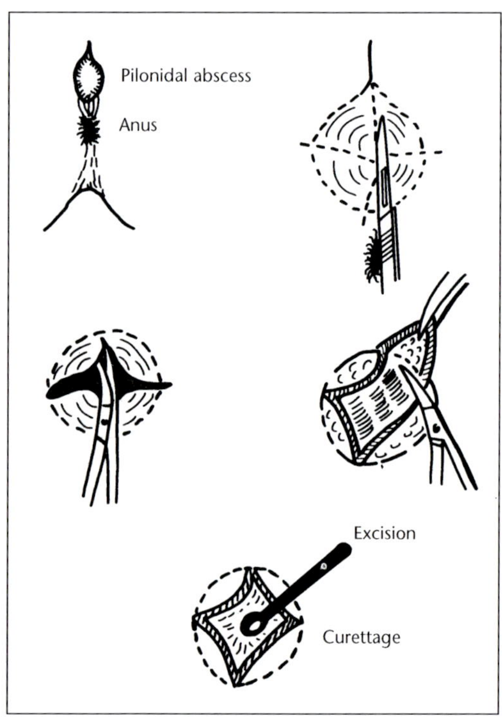

Fig. 8.6. *Cruciate incision of abscess, followed by excision of skin edges and curettage.*

to prevent the cavity from closing too quickly (Fig. 8.6). The wound is dressed with a saline mèche covered by gauze dressings. The patient can undertake the daily dressing changes himself possibly with the help of a family member. It is important that all hair in the region is shaved at the time of the incision and that the patient is instructed to keep the area free of hair thereafter. After drainage the patient should have daily sitz baths and the area should be thoroughly dried; this can be done quickly and easily with a hairdryer.

It is not necessary to give the patients antibiotics as a routine. In a Danish prospective investigation of 73 patients with acute pilonidal abscesses treated in the above mentioned manner, 58% healed primarily in the course of 10 weeks. Actuarial analysis showed a cure rate of 76% in these patients with a median observation time of 5 years. The patients that healed primarily after simple incision had significantly fewer primary or secondary sinuses than those that did not heal. The median number of days off work was 2. Other more extensive procedures are not thought to give better results, and the authors do not employ these as a routine. Other procedures consist of excision of the lesion and primary closure under antibiotic cover or excision and allowing the cavity to heal from its base.

Chronic pilonidal sinus and cyst

The treatment of a chronic pilonidal sinus or cyst depends partly on the treating surgeon's concept of the etiology. If the congenital hypothesis is favored, the lesion should be excised *in toto*. On the other hand, if the lesion is thought to be acquired, it can simply be incised and any hair and granulation tissue removed. The authors believe that this is an acquired lesion, and always try to treat it by the most simple method, described in the following.

Lord and Millar's brush method. This method of treating a chronic pilonidal sinus or cyst is extremely simple and is especially suited to local anesthesia. The method was described by Lord and Millar in 1964. The main principle involved is to remove any hair which may be found in the cavity and its secondary fistula

track, since hair is the most likely cause of recurrent infections.

The patient is treated under local anesthesia with 1% lidocaine with noradrenaline. One starts by thoroughly shaving all hair from the area to a radius of about 6 cm around the primary and secondary opening. An elliptical incision is made about 0.5–1 cm around the primary sinus (Figs. 8.7 and 8.8). The excision should extend down to the cavity where all hair is painstakingly removed with forceps or a small brush (Figs. 8.9 and 8.10). The cavity itself should not be totally excised. If a lateral sinus

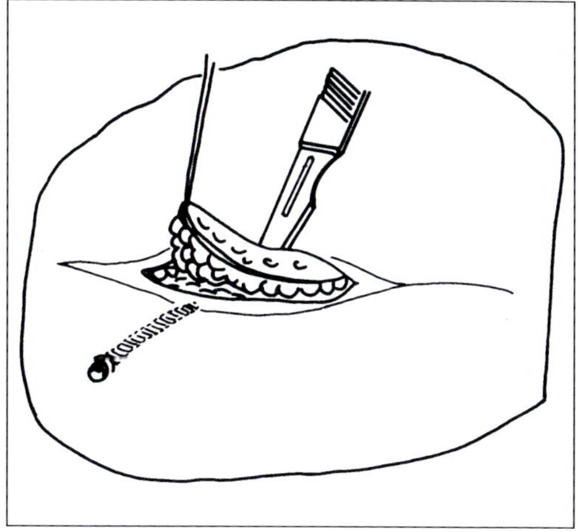

Fig. 8.7. *Illustration of the principle of the circular excision of the primary and/or secondary sinus – the method described by Lord and Millar.*

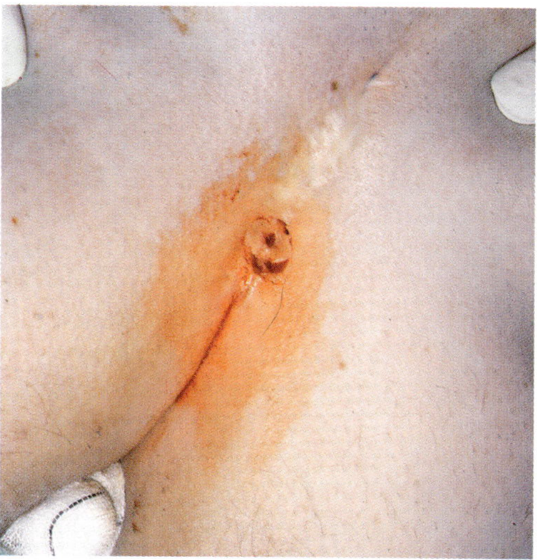

Fig. 8.8. *Circular excision of primary sinus.*

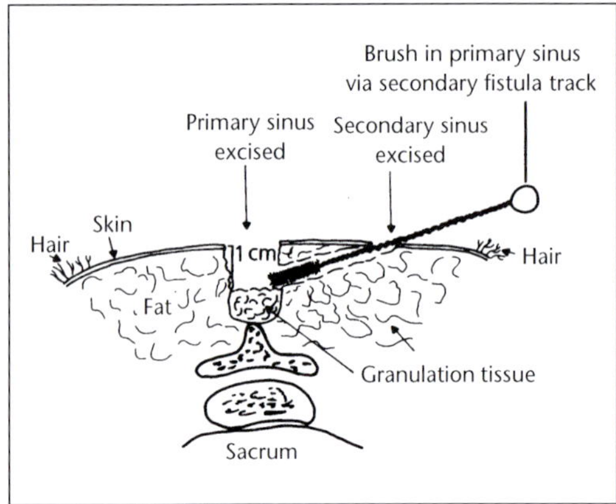

Fig. 8.9. *Schematic illustration of Lord and Millar's brush-method.*

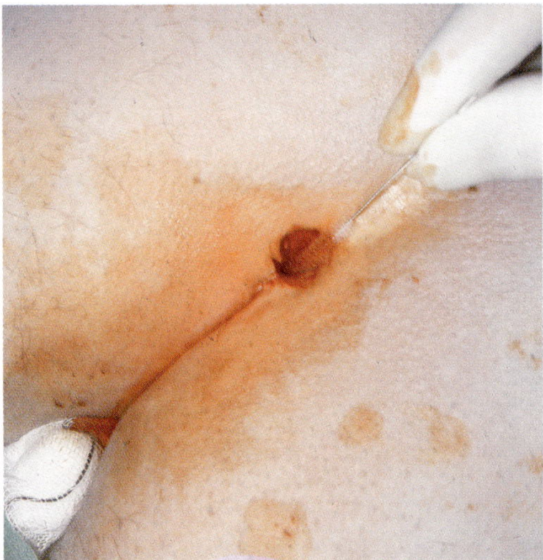

Fig. 8.10. *The excised sinus from Fig. 8.8 is brushed.*

exists it should be probed to localize fistula tracks to the cavity. This track is gently dilated and the opening excised to ensure adequate drainage. All hair from the fistula tracks must be removed either with a curette or with the brush described by Lord and Millar in 1965. We always use a small brush, which can be made by cutting the cleaning brush from a gastroscope to the required length. This type of brush is also commercially available (Cytobrush). Postoperatively the wound is dressed with a dry gauze dressing only, which is changed daily after a shower. Any bleeding from the wound edges can be controlled by diathermy coagulation. Postoperatively the patient is examined about every

2 weeks. The patient is instructed to keep the area free of hair by shaving with an electric razor every 2–3 weeks. The median length of time until healing is about 2–3 weeks. The results of Lord and Millar's brush method are promising. The patients are ambulant and able to work during the whole process, and about 90% are cured in the first instance with only about 11% developing recurrences within the first 18 months.

Total excision and secondary healing. This method is still the routine procedure for the treatment of pilonidal cysts in many centers. The actual surgical excision presents no great problems. The authors have performed the excision under local anesthesia on a few occasions, but would normally prefer to treat the patients under general anesthesia. The patient is positioned prone over a cushion or using an adjustable table, so that the sacrococcygeal region is elevated. The skin is cleaned with an antiseptic solution. An ellipsoid incision is made to include both the primary and secondary sinus openings. The incision is deepened down to the fascia, which covers the

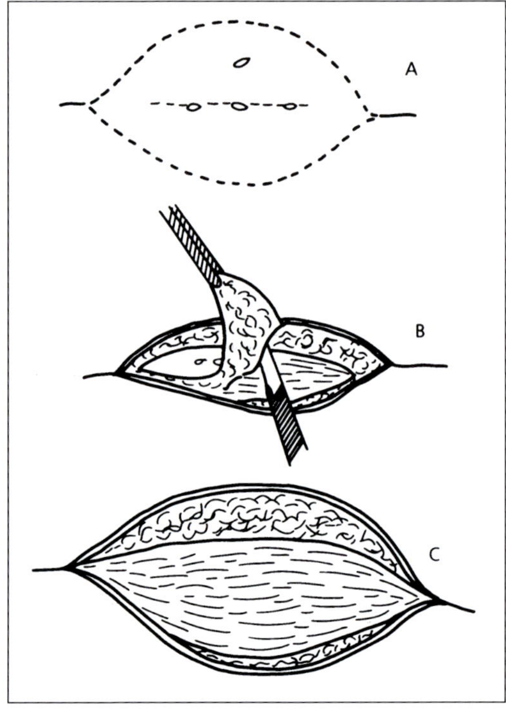

Fig. 8.11. *Complete excision of Pilonidal cyst.*

coccyx and os sacrum. Bleeding vessels are coagulated. The cyst is excised *in toto* down to the fascia in an anocranial direction (Fig. 8.11). Dye should not be injected into the sinus opening before ensuring that all the sinus tissue has been removed, since this will cause too much tissue to be removed, as dye will often be injected outside the cyst. The wound is often large and is dressed with saline mèche, which is changed on the second postoperative day and thereafter daily under general anesthesia for the first few days. The average time to healing is about 8–10 weeks, and it is normal for the patient to be hospitalized for the first 2–3 weeks depending on the size of the wound. The method is thus not especially convenient for the patient, and the recurrence rate is no better than after the Lord and Millars's brush method.

Total excision and primary closure. One of the problems with primary closure of the often large cavity after total excision of the pilonidal cyst is that if thorough hemostatis is not ensured before closure and the wound edges are completely adapted, the cavity easily becomes filled with blood and tissue fluid. If this is not the case, the wound often dehisces and infection and abscess formation rapidly occurs in the blood clot. It is, therefore, necessary to use large, deep sutures, that include the whole wall on both sides of the wound right down to the fascia. We use the suture technique shown in Fig. 8.12. After closure the wound is covered with dry, sterile dressing. Closure is carried out under relevant antibiotic cover.

Patients are hospitalized and must remain in bed for 3–5 days. The deep sutures are removed after about 7 days, and the superficial ones at about 10 days. The advantage of this method compared to secondary healing is that the patient is hospitalized for a shorter time, and discomfort is much less. The results do not seem to be better than after other methods, although a few reports have shown a very low recurrence rate after primary closure of the cavity (about 4%). Various plastic surgical methods are used to close large cavities. The potential disadvantage should infection and necrosis ensue is so great that these methods are not used as routine treatment for pilonidal cysts. They are only indicated in the few cases where extensive excision has been performed because of frequent recurrences of the condition.

Other methods. Besides the methods mentioned here, many other approaches have been described, for example, simple incision, simple incision with marsupialization of the pilonidal cyst (Fig. 8.13) in cases of severe fibrosis of the cyst wall. The authors have no experience with these methods, and since they have no

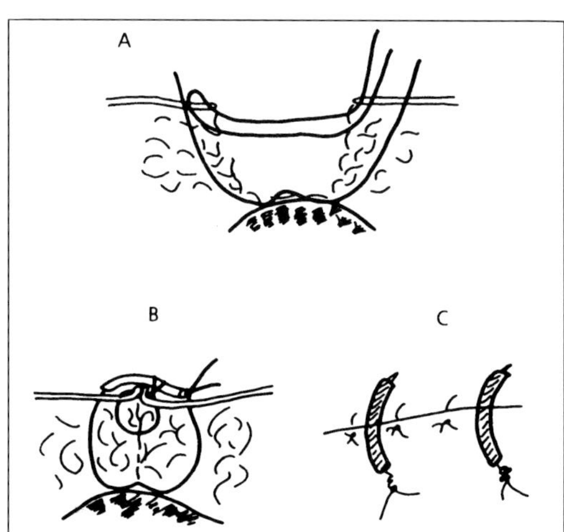

Fig. 8.12. *Suture-technique in pilonidal cyst.*

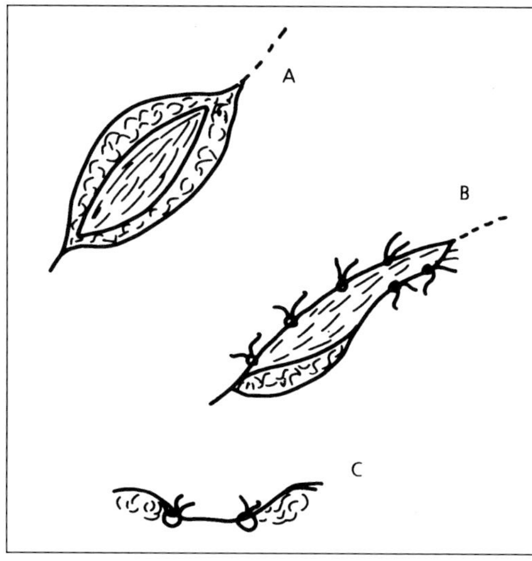

Fig. 8.13. *Simple incision of pilonidal cyst followed by marsupialisation.*

advantage over the techniques described above, they will not be discussed here.

Prophylaxis

If one believes that the condition is acquired, it is important to decrease the recurrence rate after treatment by ensuring that any hair is prevented from growing down into the skin or any sinus openings. Some have prevented hair growth by radiating the region, but because of the potential risks of skin necrosis or malignant skin changes the method cannot be recommended. We recommend simply that the patients have the region shaved about every 2–3 weeks using an electric razor. Another possibility is to suggest daily scrubbing of the area with a nail brush.

SUPPLEMENTARY READING

Bascom J. Pilonidal disease: origin from follicles of hairs and results of follicle removal as treatment. Surgery 1980; 87: 567.

Edwards MH. Pilonidal sinus: a 5-year appraisal of the Millar-Lord treatment. Br J Surg 1977; 64: 867.

Jensen SL, Nielsen OV. Lord og Millar's børstemetode ved ambulant behandling af cystis pilonidalis. Ugeskr Læger 1988; 75: 60.

Lord PH, Millar DM. Pilonidal sinus: a simple treatment. Br J Surg 1965; 52: 298.

Mansoory A, Dickson D. Z-plasty for treatment of disease of the pilonidal sinus. Surg Gynecol Obstet 1982; 155: 409.

Zimmerman C. Outpatient excision and primary closure of pilonidal cysts and sinuses. Am J Surg 1978; 640: 2.

IX. Sexually transmitted anal diseases

Sexually transmitted diseases of the anus include a large group of infections that are transmitted chiefly via anal intercourse. The diseases occur most commonly in homosexual men, but about 5–10% of women and heterosexual men also engage in anal intercourse. One should, therefore, also bear these diseases in mind when investigating women with atypical anal symptoms or anal disorders that are resistant to treatment.

FUNGAL INFECTIONS (MYCOSIS)
Definition

Perianal infections caused by various fungi. These may be transmitted by sexual contact. The commonest infection is that caused by *Candida albicans*, but infections caused by *Histoplasmosis, Blastomycosis*, and other fungi are also described.

Etiology

Asymptomatic carriers house *C. albicans* in the genitalia, but the mouth and the anorectal region are much more frequently the primary focus of infection, so the gastrointestinal tract may be regarded as a reservoir for *Candida*. The organism can be isolated from the semen of men who have been in contact with an infected partner. It is not known how commonly a perianal *Candida* infection is sexually transmitted.

Anal manifestations

Candida infections cause macerated skin with a pinkish erythema often accompanied by small satellite pustules (Fig. 9.1). Patients often complain of pruritus ani as well as pain. The infection often spreads to the gluteal cleft (Fig. 9.2) and to the groin. It may be difficult to distinguish this condition from erythrasma and pyogenic infections in this region.

Diagnosis

The diagnosis can often be made because of the typical erythema with the presence of satellite pustules. The diagnosis is confirmed by direct microscopy of biopsy material (Fig. 9.3) or by culturing a scrape from the skin.

Treatment

Candidiasis of the anal region can be treated with topical applications of creams or ointments which contain nystatin, or systemically with amphotericin B or imidazol preparation.

Sexual partners should also be treated. Recurrent chronic cases should be treated orally

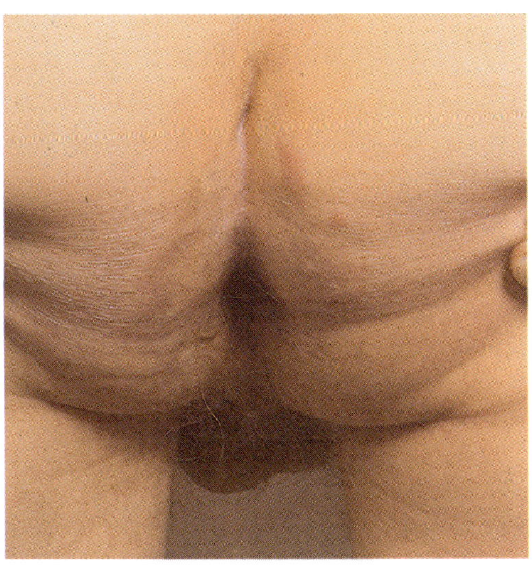

Fig. 9.1. *Perianal candidiasis.*

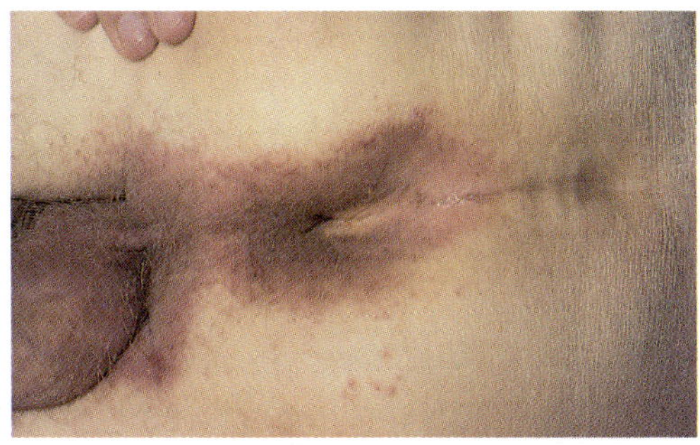

Fig. 9.2. *Anogenital candidiasis.*

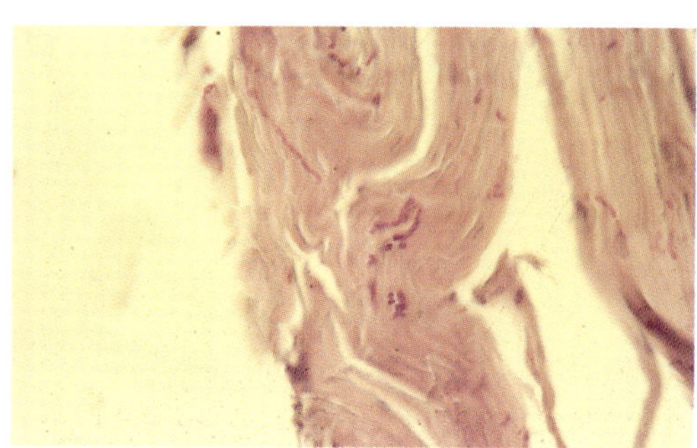

Fig. 9.3. *Histological illustration of mycosis superficialis.*

with nyastatin tablets – 500,000 I.U. q.i.d. for about 14 days. *Candida* vulvo-vaginitis can be treated with imidazol vaginal pessaries.

Patients with recurrent candidiasis, where *C. albicans* is cultured from the rectum, can be treated with ketokonazol tablets – 200 mg daily for 1–2 weeks.

PARASITIC INFECTIONS

Parasites that do not require an intermediate host and that can survive the digestive process may be transmitted by sexual contact. The following parasites have been associated with sexually transmitted anal diseases: *Giardia lamblia* and *Entamoeba histolytica*. These are particularly common in homosexual men after anal intercourse with an infected partner. Symptoms include urethritis and anal discharge of blood and mucus.

Treatment

G. lamblia infection is treated with metronidazole tablets – 400 mg t.i.d. for 7 days. Symptomatic infection with *E. histolytica* is treated with furromid tablets – 500 mg t.i.d. for 10 days plus metronidazole tablets – 800 mg t.d.s. for 5 days.

AIDS (ACQUIRED IMMUNODEFICIENCY SYNDROME)

Definition

AIDS was initially defined as a "T-cell immunodeficiency in a previously healthy adult person accompanied by opportunistic infections or Kaposi's sarcoma". This definition has come to be too restrictive since many different AIDS-related diseases exists, encompassing many different malignant tumors and lymphadenopathies.

Etiology

The serological agent of AIDS was identified in 1984 as a human retrovirus, human immunodeficiency virus type 1 (HIV-1). Three other retroviruses have since been described that apparently cause infection in the same manner as HIV-1: HTLV-1 (human T-cell leukemia/lymphoma virus type 1), HTLV-2 (human T-cell leukemia/ lymphoma virus type 2), and HIV-2 (human immunodeficiency virus type 2). HIV-1 infection occurs via sexual contact and blood-to-blood transmission. HTLV-1 and -2 are isolated from patients with lymphoproliferative disease and HIV-2 from patients with an AIDS-like disease picture. The latter three retrovirus types occur rarely in Western Europe.

Epidemiology and transmission

AIDS is not only seen in adults but also in the children of infected mothers. The risk groups are homosexuals, drug addicts (intravenous abusers), hemophilia patients (via infusion of factor VIII concentrates), and individuals from Haiti. Amongst homosexuals the passive partner in anal intercourse as well as those who have many partners are at the greatest risk.

Anal manifestations

The anal manifestations of AIDS are secondary to the immune defect and are due in most cases to opportunistic infections. Viral and fungal infections, including herpes simplex, are particularly common in the perianal area. Kaposi's sarcoma may also be seen perianally (Fig. 9.4). The lesion presents as a dark blue or purple-brown plaque or lump that may ulcerate. The patients may also have an accompanying carcinoma of the anal canal.

Diagnosis

Diagnosis is made on the basis of patient history, clinical findings, and laboratory data. There must be signs of immunodeficiency; there is marked lymphopenia, a diminished

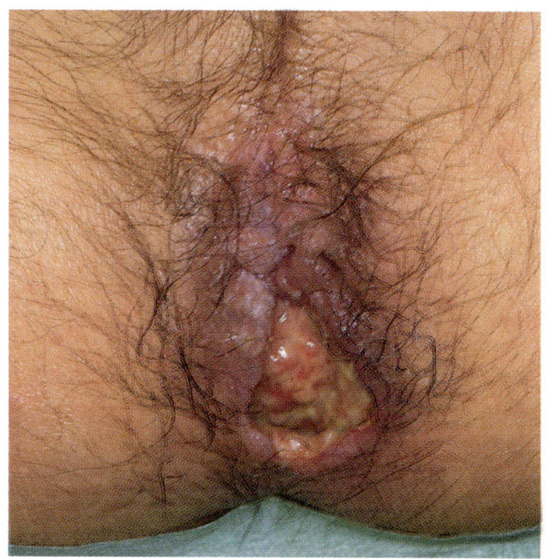

Fig. 9.4. *Perianal Kaposi sarcoma in a patient with AIDS.*

number of T-cells, and the helper/suppressor ratio is especially depressed. However, it is not sufficient to demonstrate immunological changes since these are seen in various viral diseases as well as during treatment with most immunosuppressive drugs. The clinical picture must also conform to diagnose AIDS.

Treatment

There is no specific treatment for AIDS. The antiviral drug azidothymidin (Retrovir®) has an effect on the survival of AIDS patients, but the treatment is accompanied by severe side-effects. The anal manifestations can be treated symptomatically according to the nature of the lesion. Herpes simplex can be treated with aciclovir; *Candida* infection with amphotericin B. Malignant tumors of the anal region can be removed in the usual way (see chapter 13).

GONORRHEA
Definition

A bacterial infection caused by *Neisseria gonorrhoeae*, a Gram-negative diplococcus. The infection is nearly always transmitted by sexual contact. The physical signs, symptoms, and complications as well as the disease's natural history are very different in men and women. Many cases are asymptomatic (15–60%). The main symptom in men is urethritis and in women acute

salphingitis or vaginal discharge. Anorectal symptoms and signs occur in both sexes.

Anal manifestations

Anorectal colonization occurs rarely. The incidence of positive cultures from the region correlates with the frequency of anal intercourse. If it is suspected that the patient has been in contact with a gonorrhea-infected partner, cultures should always be taken from the anal canal, rectum, cervix, urethra, and pharynx. The symptoms arising from the anorectal region are not diagnostic for gonorrhea. These may be symptoms of proctitis, an ulcer or possibly an abscess of the anal canal or the perianal region. Perianal dermatitis with edema and multiple fissures or erosion is often seen.

Diagnosis

Diagnosis is made by demonstration of diplococci on Gram-staining or on culture. It is not sufficient to culture from the urethra. The cervix, rectum, and pharynx should also be swabbed. In culturing from the rectum media should be used which contain antimicrobial substances that inhibit all other organisms except *Neisseria*. Cultures from perianal lesions or from anorectal abscesses are often negative. There are many serological tests under investigation, but none has the necessary sensitivity and specificity. Biopsies of lesions, such as perianal lesions, will typically show an acute or subacute inflammatory reaction dominated by polymorphonuclear leucocytes. The only objective sign will often be an abscess or an ulcer in the anal region. In patients at high risk of infection (homosexuals and prostitutes) who have ulcers or abscesses in the anal region, cultures should be taken from the urethra, cervix, and rectum.

Treatment
Penicillin and probenecid

Uncomplicated gonorrhea can be treated as follows:

1. 1.4 g pivampicillin together with 1 g probenecid orally as a single dose.

2. 3.5 g ampicillin together with 1 g probenecid as a single dose.

3. 3 g amoxicillin together with 1 g probenecid orally as a single dose.

Tetracylin and erythromycin

Tetracyclin and erythromycin in doses of 500 mg q.i.d. for 5 days can also be used. Tetracyclin, however, is not often used because of its potentially serious side-effects.

Spectinomycin and cefuroxim

Gonorrhea caused by penicillinase-producing gonococci is treated with spectinomycin (2 g i.m.) or cefuroxim (1.5 g i.m.) preceded by 1 g probenecid given orally 30 minutes before injection.

Sulfa preparations

In cases of penicillin allergy sulfamethoxacol and trimetoprim tablets – 400 mg + 80 mg in two doses at 8 hour intervals – should be used. The same treatment should also be used if there is suspicion of simultaneous syphilis in order not to mask this.

SYPHILIS

Syphilis is caused by *Treponema pallidum.* Transmission occurs in most cases by sexual contact, but infection can occur in other ways. The early stages of the disease are usually characterized by mucocutaneous lesions which often ulcerate.

Epidemiology

Syphilis occurs universally and is seen most frequently in the sexually active age groups. The disease is more common in men than in women as it is often spread by homosexual contact. It is difficult to trace the source of infection especially amongst homosexuals, since homosexual partners are often anonymous and one may be dealing with multiple partners. The incidence is thought to be rising, but reliable statistics are not available.

Transmission

Transmission of *T. pallidum* occurs by direct contact and may be seen after blood transfusion or oral intake of menstrual blood or vaginal secretions. In the hospital environment pathologists, surgeons, and laboratory assistants are at high risk.

Classification

The disease is divided into a primary stage (chancre), a secondary stage (mucocutaneous lesions with lymphadenopathy and other organ involvement), a latent stage (positive serological test without objective or subjective signs of the disease), and a tertiary stage (involving the CNS, cardiovascular system, skin, and internal organs). For the purposes of treatment the disease is often classified in another way, namely according to its duration: 1) early syphilis (primary and secondary syphilis) of less than 1 year's duration; 2) latent syphilis (seropositive cases without symptoms or objective manifestations); and 3) late syphilis of more than 1 year's duration with manifest disease and benign involvement (gumma) of the cardiovascular system and CNS.

Early syphilis

Primary syphilis. The primary lesion of syphilis is the chancre which may also be found perianally. The chancre is initially an inflammatory papilla which later erodes forming a crater producing an exudate. The chancre is most often localized to the genitalia, but it may be seen extragenitally in the anus or the rectum with or

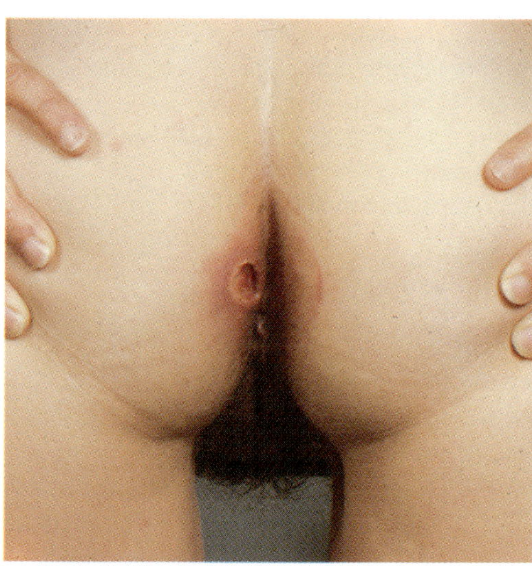

Fig. 9.5. *Syphilitic chancre at the anal margin – at first mistaken for a fissure.*

without accompanying proctitis and inguinal lymphadenitis. It is often painful, especially when secondary bacterial infection is present.

In the anal region or the rectum the chancre often resembles a polyp, a malignant tumor or a fissure of the anal edge and is often primarily misdiagnosed as such (Fig. 9.5). The lesion is most often situated in the posterior midline. The symptoms may be pain at defecation or severe nocturnal pain as seen with a common anal fissure. The presence of bilateral lymphadenopathy should aid diagnosis since lymphadenopathy is extremely rare with other lesions.

Secondary syphilis. This occurs about 2–6 months after primary infection. The earliest lesion is a macula which arises 3–6 weeks after the chancre. This appears together with pustules and papules. The macula presents as an oval spot which does not desquamate or itch. The lymph nodes are also usually involved. About 3–5 weeks after formation of the macula, papilloma elements with well-defined macerated surfaces may appear. These papules are called condylomata lata (Figs. 9.6 and 9.7) and should not be confused with condylomata acuminata.

Late syphilis

Tertiary syphilis. The lesions are seen many years after the primary infection and represent an

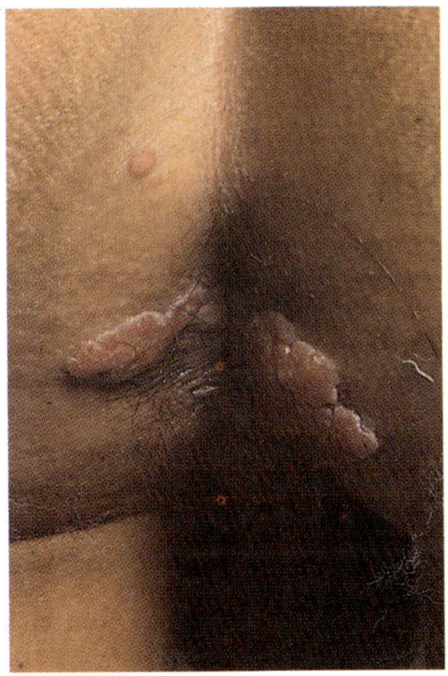

Fig. 9.7. *Perianal condylomata lata.*

inflammatory response to *T. pallidum*. These are superficial nodular gummata in the skin several millimeters in size and brown-red in color. The lesions may ulcerate and become scabbed.

Diagnosis

The diagnosis is made by serological tests, darkground microscopy, and possibly demonstration of the presence of *T. pallidum* in biopsies (Fig. 9.8). The serological test is based on the formation of two types of antibodies: 1) antireponemale and 2) antipoidale (Wassermann reaction [WR] etc.). The reader is referred to textbooks on venereology.

Treatment

Penicillin is still the drug of choice in the treatment of syphilis, irrespective of the stage of the

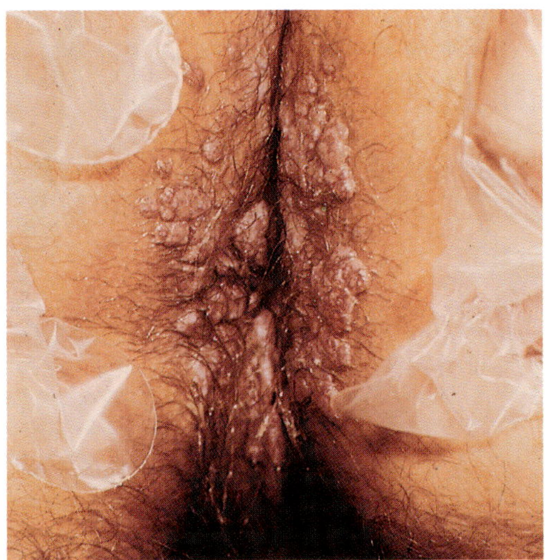

Fig. 9.6. *Perianal papillomatous condylomata lata.*

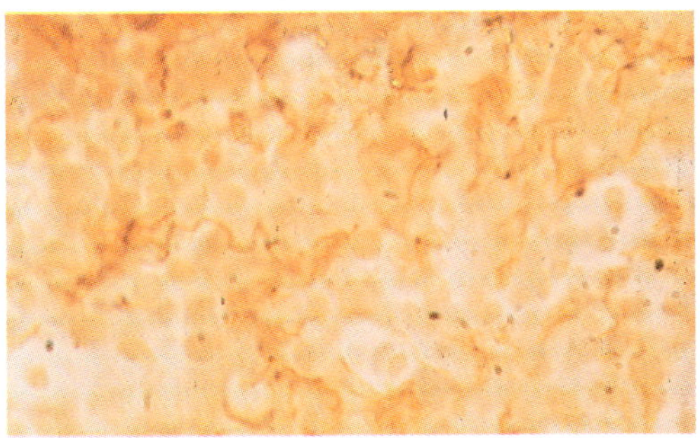

Fig. 9.8. *Microscopic view of treponema pallidum from the anal region.*

disease. For syphilis of less than 1 year's duration a dose of 2.4 million I.U. is given as a single i.m. injection. Prokainpenicillin may also be used in a dosage of 600,000 I.U. daily for 8 days. For syphilis that has been present for more than 1 year, penicillin G is used in a total dosage of 7.2 million I.U. (2.4 ml per week for 3 weeks i.m.) or prokainpenicillin in a total dosage of 9 million I.U. i.m. (600,000 I.U. daily for 15 days). For patients with penicillin allergy tetracyclin (500 mg q.i.d. for 15 days), erythromycin (500 mg q.i.d. for 15 days), or a cefalosporin (e.g., cephaloridin 500 to 1000 mg daily for 6 to 10 days) can be used for both early and late syphilis.

LYMPHOGRANULOMA VENEREUM
Definition

A sexually transmitted disease caused by *Chlamydia* and involving the lymphatic system. The location of the disease depends amongst other things upon the patient's sexual habits and is found most commonly in the genitalia. It may also however, occur in the anorectal region and in the pharynx. This condition is of differential diagnostic significance in conditions that cause proctocolitis and anorectal strictures.

Epidemiology

This rarely occurs in Denmark and the Western world, and its incidence is thought to be falling. Is seen most frequently in the second to fourth decade, especially in patients of the lower social classes.

Etiology

The disease is caused by *Chlamydia trachomatis* types I, II, and III.

Anal manifestations

The primary lesion is a 4–8 mm soft, erythematous lesion that heals spontaneously in the course of a few days. A few weeks after the appearance of this lesion, tender lymphadenitis occurs in the inguinal regions. Lymph node swelling appears in the course of a few days. Direct infection of the anal canal in men and of the vagina in women may cause acute symptoms in the anorectal region that include anal

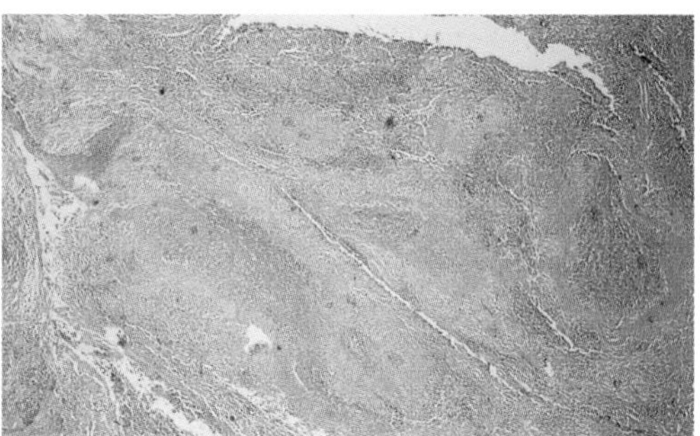

Fig. 9.9. *Typical histological example of lymphogranuloma venereum.*

and rectal pain, tenesmus, and a mucous sanguinous discharge from the anus.

On rectoscopy one finds the typical signs of proctocolitis. Anorectal stricture formation can be a late complication.

Diagnosis

Diagnosis is made from the typical histological picture (Fig. 9.9) or by isolating and identifying the organisms, Frei's test or a complement fixation test. The latter test usually becomes positive 4 weeks after infection. A quadrupling of the titer is considered to be the best proof of the disease. The reaction is, however, not specific.

Treatment

Standard treatment is erythromycin or tetracyclin at a dose of 500 mg q.i.d. for 3 weeks. The treatment is effective in the acute phase, but not at the stage of lymphadenitis. Penicillin is ineffective.

CHANCROID
(ULCUS MOLLE, SOFT CHANCRE)
Definition

An acute localized infectious disease caused by the Gram-negative bacterium *Haemophilus ducreyi*. It is characterized by a necrotic ulceration at the site of infection. This is accompanied by regional lymphadenopathy.

Etiology

The disease is caused by *H. ducreyi* which is a short, non-spore-bearing, non-acid, Gram-negative bacterium. The incubation period is about 1 week.

Epidemiology

This condition is very rare in the Western world. It is most common in undeveloped countries. It is usually sexually transmitted. It is seen more commonly in men than in women. Female prostitutes are often asymptomatic carriers.

Anal manifestations

The disease begins with a papule that develops into a pustule and rapidly becomes a small, soft, well-defined ulceration. The sore is often covered by a yellow-grey exudate and bleeds easily when touched. The sore is often tender and painful as opposed to the lesions of syphilis and is more commonly seen on the genitalia than in the anus. It may be accompanied by tender, fluctuating regional lymph nodes. These may suppurate if the disease is untreated.

Diagnosis

An intracutaneous test (Ducrey) is positive after a few weeks. The reaction is, however, not specific. There is a characteristic histological picture.

Treatment

Sulfonamides, tetracyclin, erythromycin or lincomycin in usual doses. Penicillin has no effect on the disease.

HERPES ANALIS
Definition

An infection caused by one of the two herpes simplex viruses, type 1 (HSV-1) or type 2 (HSV-2). The perianal infection is most commonly caused by HSV-2. The disease is characterized by recurrent infections. The infection is usually transmitted by sexual activity.

Epidemiology

The incidence of herpes infections in the perianal region has increased markedly in recent decades together with genital herpes infections. Genital herpes infection is the commonest venereal disease of the middle-class in USA,

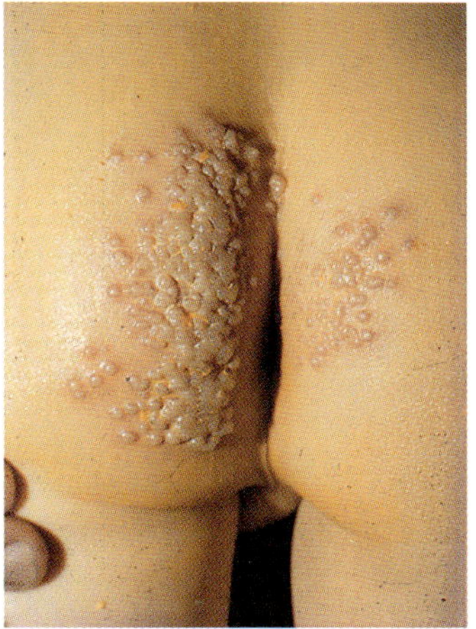

Fig. 9.10. *Typical herpes vesicula in the perianal region.*

and anal herpes occurs frequently in homosexual patients. It is characteristic for the virus to recur after the primary infection. The recurrent disease is usually milder and of shorter duration than the primary disease. The virus is found latent in the regional nerve ganglia after primary infection, and it is from here that the recurrences arise.

Anal manifestations

Primary herpes in the perianal region or the anal canal is often characterized by severe pain, discharge, tenesmus, and secondary obstipation. On physical examination the region around the anus is red, edematous, and covered with mucus. There are painful erosions or fissures and possibly groups of blisters (Fig. 9.10). The condition should not be confused with the lumbosacral herpes simplex infection situated on the buttocks (paragenital herpes), where the pain often radiates to the lower extremities. The anal lesions are often accompanied by tender lymph nodes in the inguinal regions.

In cases of recurrence the symptoms are different and may be mainly pruritus ani or symptoms of a fissure. The severity of infection and frequency of recurrence varies greatly from patient to patient. Some may have multiple recurrences in the course of a single month.

Diagnosis

It is important to recognize the disease when treating patients with recurrent anal symptoms,

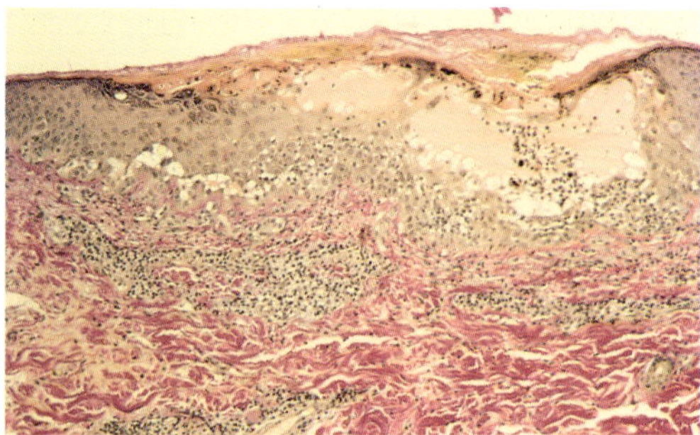

Fig. 9.11. *Typical histological example of herpes vesicula.*

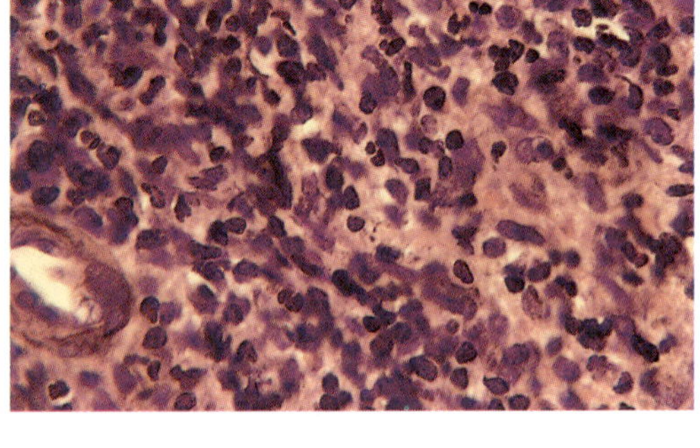

Fig. 9.12. *Gram negative Donovan bodies in a biopsy from a patient with granuloma inguinale.*

especially if they complain of recurrent painful ulcerations with tender inguinal lymph nodes. The diagnosis is made by visualizing the herpes virus in the lesions (Fig. 9.11). The diagnosis may also be made serologically, but this method is seldom used.

Treatment

The primary attack is treated with oral aciclovir (Zovirax® 200 mg × 5 times a day for about 10 days), while the less effective topical aciclovir treatment can be employed for recurrences (Zovirax® cream 5% 5 times daily for about a week). Recurrences may also be treated with oral aciclovir and if the patient has the tablets available at home and starts the treatment as soon as the prodromal symptoms appear the prognosis is better than if treatment is delayed. If the recurrences are exceptionally frequent and in immunosuppressed patients, intravenous treatment with aciclovir can be considered. In these groups of patients prophylactic aciclovir treatment can be considered.

GRANULOMA INGUINALE
Definition

A skin disease that can be spread both by sexual and non-sexual contact. It is a progressively ulcerating and granulomatous skin disease caused by *Calymmatobacterium granulomatis.*

Epidemiology

This occurs rarely. Is found world-wide, but occurs most frequently as an epidemic disease in tropical and sub-tropical regions. The reason for suspecting that the disease it sexually trans-

mitted, is that it is located in genitalia and in the perianal area in homosexuals. The disease also occurs, however, amongst children, who are not sexually active.

Anal manifestations

The primary lesion is a subcutaneous papule or lump which ulcerates within a few days. It is most commonly located on the penis or scrotum in men and the labia minora in women. From here the disease spreads directly or by autoinoculation to the perianal region. Secondary bacterial infection may cause necrotic lesions with severe tissue destruction.

Diagnosis

Diagnosis is made from Wright or Giemsastained biopsies where diagnostic Gramnegative Donovan bodies can be demonstrated (Fig. 9.12). The diagnosis can, however, be made from the clinical picture if one is aware of the disease. Distinguishing this from other sexually transmitted diseases or an ulcerating squamous cell carcinoma in the perianal region may be difficult in atypical cases.

Treatment

Oral tetracyclin 2 g daily for 10–15 days.

ANORECTAL LESIONS IN HOMOSEXUALS (THE GAY BOWEL SYNDROME)

Anal intercourse is performed by homosexual men as well as an unknown number of heterosexual persons. A plethora of microorganisms can be transmitted by this type of intercourse. The mucosa of the rectum is more vulnerable than that of the vagina, and mucosal lesions occur more easily during anal intercourse, which enhances the transfer of microorganisms, such as hepatitis-B virus, cytomegalovirus (CMV), and probably the different forms of retrovirus (HIV-1 and -2 as well as HTLV-1 and -2). Enteric microorganisms may be spread by sexual contact involving the anorectum, mouth, and hands.

The gay bowel syndrome encompasses many different diseases and should not be considered as an actual syndrome. Many infections of gay bowel syndrome are asymptomatic, and are not diagnosed unless microbiological examinations are carried out. This together with the fact that it may be difficult to obtain accurate information on the patients' sexual habits results in many of the conditions being overlooked. When treating homosexuals it is important to realize that these patients often have coexisting infections, and one cannot be sure that an organism which has been isolated is the cause of the current symptoms and signs.

Gonorrhea is the commonest sexually transmitted disease amongst homosexuals. The anorectum is the only site of infection in about 40% of these patients, most of whom are asymptomatic. In about half of the patients diagnosis is made by Gram-staining of a rectal smear making it necessary to take cultures in order not to overlook the disease.

Syphilis is seen almost exclusively in its earlier stages. By far the majority of newly diagnosed cases of syphilis are seen amongst homosexual men (about 80% of cases). The anorectal chancre might be atypical and is easily overlooked by both the patient and the doctor. It might resemble an anal fissure. Perianally the secondary syphilitic lesions appear as condylomata lata and should not be confused with condylomata acuminata. The disease will often be present in an early latent phase without accompanying symptoms. For this reason homosexuals with many unknown contacts should be screened regularly with serological tests. Darkground microscopy can be used to identify *T. pallidum* from primary and secondary lesions.

Chlamydia trachomatis (non-LGV) may cause an asymptomatic rectal infection or mild proctitis as opposed to infections with LGV-serotypes which may present with serious infections with perianal abscesses, fistulae, and rectal strictures, so that the disease is often confused with Crohn's disease. Diagnosis is made primarily by isolating the organism, sometimes together with serological tests. Infections with *G. lamblia* and *E. histolytica* are seen endemically amongst homosexuals. About 20% of homosexuals who are under treatment for venereal disease excrete cysts. Infections are most often asymptomatic. It is controversial whether asymptomatic homosexual men who excrete cysts should be treated. Incidence of condyloma and herpes infections of the anal region is increasing massively amongst homosexuals, and anal condylomata are about 10 times more common than penile warts in homosexuals. About 60% of homosexuals with anal condylomata also have condylomata in the rectum. A strong relationship has been shown between the human papilloma virus and cervical neoplasms in women and squamous cell carcinoma of the anal region in homosexual men. Herpes infection is a common cause of proctitis in homosexuals. In the presence of AIDS, herpes simplex infections can be extremely serious with widespread ulcerating perianal lesions, fever, neurological symptoms, and constipation. Aciclovir intravenously or orally often has a good effect.

Infections with *Enterobius vermicularis, Pthirus pubis* and *Sarcoptes scabiei* may often be the cause of pruritus ani in homosexual men and searching for eggs is obligatory in the examination of homosexual patients. It is easy to overlook these infections, so repeated examination of the feces is necessary.

Investigation of a patient with gay bowel syndrome is ideally undertaken in cooperation with a proctologist, an expert in infectious diseases, a venereologist, and a pathologist.

HIV-1 infections occur endemically amongst homosexuals. The infection is chronic, and the virus can be isolated from the blood, semen, and sputum of infected individuals. The strongly increasing incidence of HIV-1 infections in homosexuals should lead one to consider all homosexuals as potentially infected.

The final stage of HIV-1 infection is AIDS. The bowel is one of the most common sites for the disease's tumors and opportunistic infections. Kaposi's sarcoma is the first sign of the disease in about 21% of the patients. There can be severe involvement of skin, mucous membrane, lymph nodes, and organs with lesions from the pharynx to the rectum. The disease may present in the bowel as small flat lesions which are visible on endoscopy, or as large nodular tumors which may also be seen on X-ray examination.

Fungal infection is one of the commonest opportunistic infections in AIDS patients. Candidiasis occurs especially in the oral cavity and the esophagus. Diarrehea in AIDS patients may be caused by infection with *Mycobacterium avium intracellulare, Cryptosporidium* (protozoan), or CMV.

It is important that the necessary facilities are available to investigate adequately homosexual patients for the infections under discussion, since such infections in AIDS patients will run a much more serious course than in otherwise healthy patients.

SUPPLEMENTARY READING

Bolling, DR. Prevalence, goals and complications of heterosexual anal intercourse in a gynecologic population. J. Reprod Med 1977; 19: 120.

Catterall RD. Sexually transmitted diseases of the anus and rectum. Clin Gastroenterol 1975; 4: 659.

DeVita (moderator). Drug development for acquired immunodeficiency syndrome. Ann Intern Med 1987; 106: 656.

Drusin LM. The role of surgery in primary syphilis of the anus. Ann Surg 1976; 184: 65.

Goodell SE, Quinn TC, Mkrtichian E. Herpes simplex proctitis in homosexual men. Clinical, sigmoideoscopic and histopathological features. N Engl Med J 1983; 83: 582.

Jacobs E. Anal infections caused by herpes simplex virus. Dis Colon Rectum 1976; 19: 151.

Marino AWM, Mancini HW. Anal eroticism. Surg Clin North Am 1978; 58: 513.

Minnesota State Department of Health Venereal Disease Control Program: Recommended Treatment Schedule for Sexually Transmitted Diseases, June, 1975.

Muggia FM, Lonberg M. Kaposi's sarcoma and AIDS. Clin North AM 1986; 70: 665.

Owen RL, Hill JL. Rectal and pharyngeal gonorrhea in homosexual men. JAMA 1972; 220: 1315.

Schachter J. Chlamydial infections. N Engl J Med 1978; 298: 428.

Schachter J. Confirmatory serodiagnosis of lymphogranuloma venereum proctitis may yield false-positive results due to other chlamydial infections of the rectum. Sex Transm Dis 1981; 8: 26.

Sohn N, Robilotti JG. The gay bowel syndrome, a review of colonic and rectal conditions in 200 male homosexuals. Am J. Gastroenterol 1977; 67: 478.

Zellner SR, Trudeau WL. Anorectal gonorrhea presenting as ulcerative proctitis. South Med J 1973; 66: 706.

X. Condylomata acuminata

Definition

Condylomata acuminata is an infectious disease caused by human papilloma virus. It is localized to mucosa and/or skin of the genitalia and anorectum, and is transmitted chiefly by sexual contact.

Etiology

As with other warts these condylomata are caused by human papilloma virus (HPV). Judging from electron microscopical examinations there does not seem to be any morphological difference between the virus causing anogenital condylomata and that causing warts. On the other hand, there is perhaps a difference in antigenicity. In serological and molecular hybridization techniques up to 40 different types of HPV virus can be demonstrated. It is considered proven that the type of HPV virus determines at least partially location and appearance of the wart or condyloma. Different viruses can cause the formation of the same kind of wart or condyloma. Type 6, 10, 11, 16 and 18 cause condylomata acuminata, and types 6, 10, and 11 in particular are found in the condylomata of the anogenital region. It should be noted that some types (types 16 and 18) are associated with malignant tumors.

Epidemiology

There is clear evidence that condylomata acuminata are sexually transmitted. It has been shown in an investigation of 332 patients with genital condylomata that 64% of their sexual contacts developed condylomata after an incubation period of up to 2 years. Genital condylomata occur much more frequently than anorectal condylomata. About 10% of men and about 20% of women with genital condylomata will, however, develop anorectal condylomata as well. Most of these practice anal intercourse. Patients practicing oral sex will not infrequently have condylomata in the oral cavity.

Anal condylomata occur 5 to 10 times more frequently amongst homosexuals than amongst others. Homosexual patients with anal condylomata often have various other sexually transmitted diseases in the region and should be thoroughly examined for these.

Condylomata in children

Anogenital condylomata are extremely rare in children (Figs. 10.1 and 10.2). Anogenital condylomata in infants may be due to contamination during birth from an infected birth canal, but condylomata in children older than 12–18 months should arouse the suspicion of sexual abuse of the child, and a careful examination should be instituted. It is possible with DNA-hybridization techniques to demonstrate the type of virus causing condylomata in children. If it is type 6, 10 or 11 the suspicion of sexual abuse is strengthened.

Histology

Condylomata usually appear as papillary lesions with parakeratosis and only slight thickening of the corneal layer (Fig. 10.3). There is frequently a marked acanthosis and papillomatosis and the blood vessels in the superficial corneal layer and in the papillae are dilated. There is often an inflammatory infiltrate present, and this is usually due to irritative changes.

Anal manifestations

The typical perianal condyloma is a skin tumor with a lobular, cauliflower-like appearance. The appearance may, however, be very variable. The tumors frequently bleed easily and are most often multiple (Figs. 10.4 and 10.5). They can present as massive cauliflower-like tumor masses that completely cover the anus (Fig. 10.6). The condylomata often extend up into the anal canal and even into the rectum.

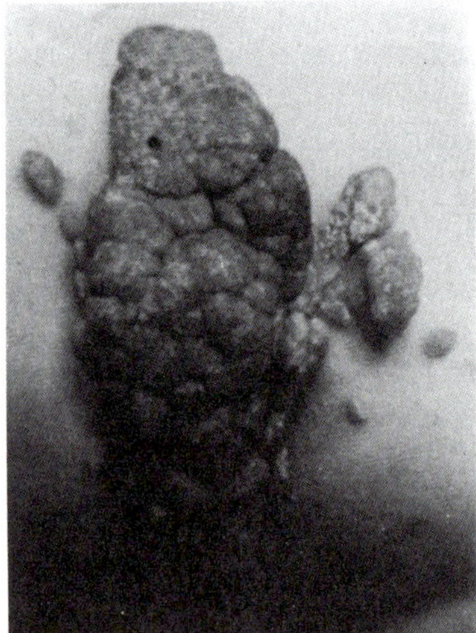

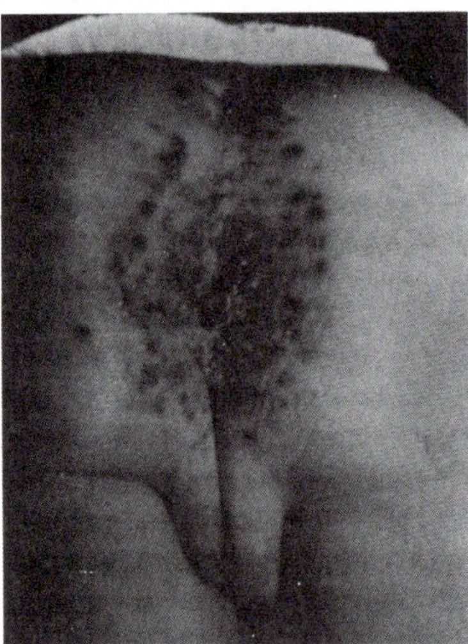

Fig. 10.1. *Large condylomata in a three-year old girl covering the anus completely – before and after surgical excision.*

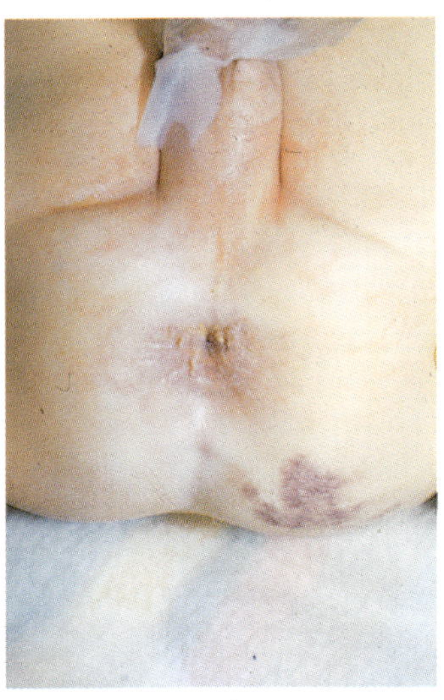

Fig. 10.2. *Condylomata acuminata in the perianal region of a two-year-old child who had been sexually abused. Notice hematoma on the left buttock due to violence.*

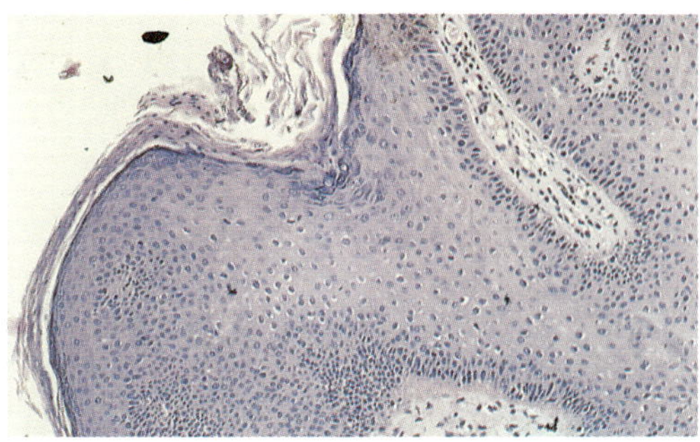

Fig. 10.3. *Histological example of condylomata.*

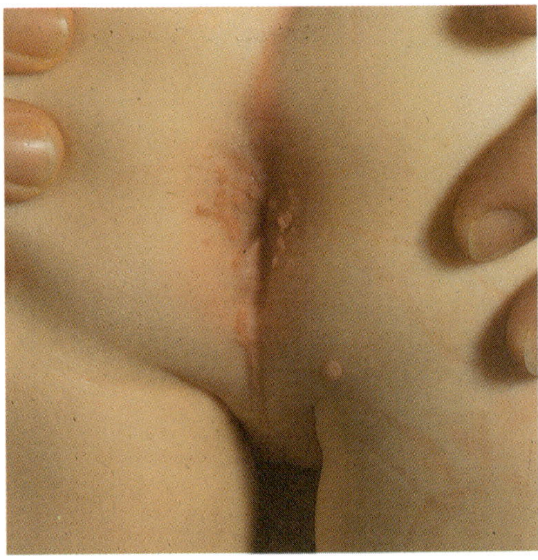

Fig. 10.4. *Multiple papillomatous condylomata in the perianal region of an eight-year-old girl who had bathed with infected adult women.*

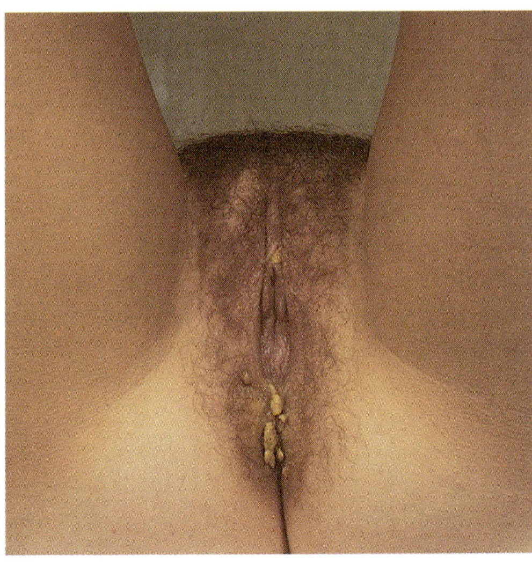

Fig. 10.5. *Multiple perianal papillomatous condylomata in an adult woman.*

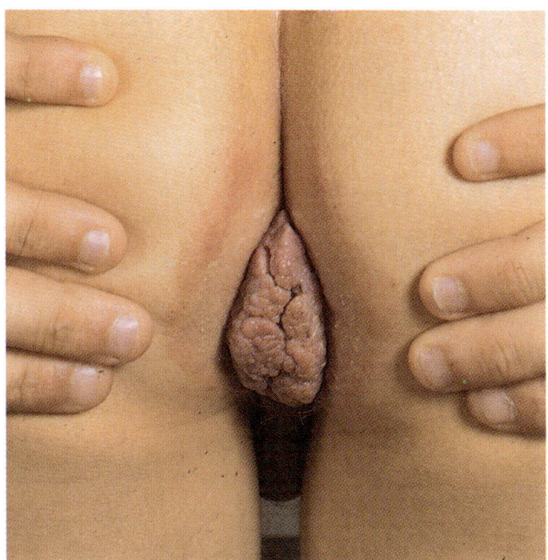

Fig. 10.6. *Large condylomata covering the whole anal region in an adult person.*

Diagnosis

The diagnosis is made primarily from the typical cauliflower-like appearance of the perianal tumors, which may cause bleeding, and is confirmed by histological examination. The WR should be negative. The examination should consist of a meticulous anorectoscopy and a thorough examination of the genitalia including the urethra orifice and the oral cavity. In addition, all patients presenting genital condylomata should also undergo a thorough examination of the anus and rectum. Not all HPV infections are visible, and it is therefore a good idea to swab the genitalia and perianal region with 3% chloroacetic acid solution. This solution makes invisible lesions appear as white elements in the skin allowing them to be treated. Gynecological examination should obviously be carried out in women. A colposcope can be used to search the region for condylomata. In patients presenting with recurrent attacks cystoscopy should be considered, since condylomata may also be found in the mucosa of the bladder, and these may be responsible for maintaining the infection. The place of routine cystoscopy in patients with anogenital condylomata is not clear. A patient's sexual partners should also be examined for the presence of condylomata.

Differential diagnoses

Perianal condylomata should first and foremost be distinguished from condylomata lata and Bowenoid papules. Atypical carcinoma *in situ* lesions may also be misdiagnosed as condylomata.

Treatment

Podophyllin resin

Condylomata are traditionally treated by the weekly application of podophyllin in concentrations of 10–25%. The podophyllin is thoroughly washed away after 4–5 hours in order to avoid a violent tissue reaction. Podophyllin should be applied to the actual condylomata, and not to the surrounding normal skin. The treatment is repeated at weekly intervals, until the condylomata disappear. When dealing with very large condyloma masses only a small area should be treated in the first instance as podophyllin is very toxic and each patient reacts differently to the treatment. Patients should not be allowed to use podophyllin on their own since they will tend to overtreat themselves causing large wounds with secondary infection resulting in great discomfort. In addition, as already mentioned, the substance is highly toxic with serious side effects. The treatment should not continue for more than 12 weeks, as the substance is potentially oncogenic. Caution is advised when using podophyllin to treat infants. Condylomata with a certain amount of resistance to podophyllin appear to have developed. This fact together with the high recurrence rate after podophyllin treatment lead the authors to recommend that first attacks of condylomata be treated by simple surgical excision under local anesthesia. Young children should, however, be treated under general anesthesia in order to obviate the psychological trauma of the local treatment. A Danish controlled investigation has shown that the cure rate after surgical excision of perianal condylomata was larger and the recurrence rate significantly less (than after podophyllin treatment) (Fig. 10.7). Nowadays first episode condylomata is treated by topical application of a podophyllotoxin cream administered by the patient.

Other methods of treatment are cryosurgery, CO_2 laser, argon beaming and electrocoagulation.

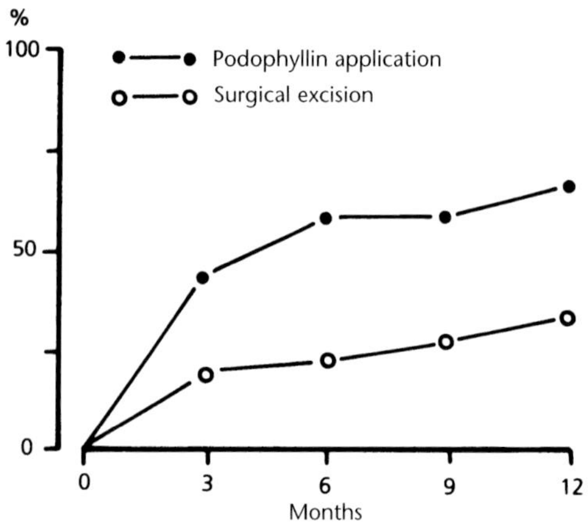

Fig. 10.7. *Recurrence-frequency of condylomata treated with simple surgical excision or podophyllin in alcohol (20%) A significantly better effect is obtained by simple surgical excision.*

These methods do not show any advantages in controlled investigations over surgical excision. This is also the case when they are used for treating recurrent anal condylomata. Condylomata in the rectum may also be removed during an outpatient treatment procedure by using McGivney's apparatus where rubber bands are placed on the mucosa beneath the condyloma. The condyloma is cut away without any bleeding (Fig. 10.8).

Trichloroacetic acid

Trichloroacetic acid (75%) can be used for small perianal condylomata. It is applied with a cotton wool swab and the lesion disappears within a few days. The treatment can be repeated after 2–3 weeks. Recurrences occur frequently.

Fluorouracil (5-FU)

Attempts have been made to use systemic cytostatic agents such as bleomycin, methotrexat, and 5-FU in the treatment of condylomata, but because of their potential side-effects systemic use should be avoided.

5-FU is an antimetabolite which causes necrosis and destruction of tissue. It can be applied locally daily until erythema or blisters develop, usually after 8–12 days. 5-FU is especially suitable for treating small multiple polyps in the vagina.

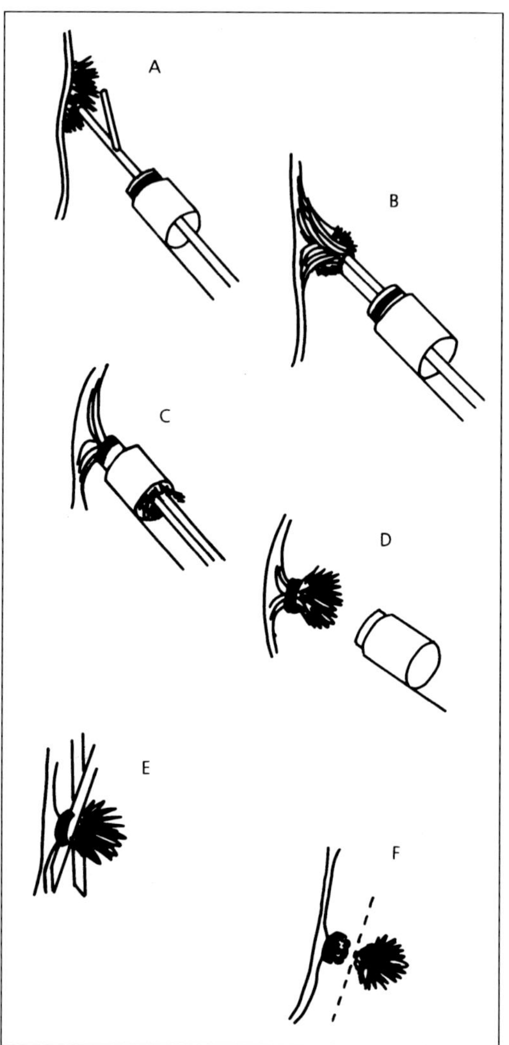

Fig. 10.8. *Removal of condylomata in the rectum with McGivney ligator and surgical excision.*

Interferon

Alpha- and beta-interferon have been used both topically and systemically in the treatment of genital condylomata. Experience is lacking in the use of interferon for anal condylomata. The advantage with systemic interferon is that both visible and invisible papilloma virus lesions are treated.

Simple surgical excision

Figs. 10.9–10.11 show the principles of simple surgical excision. After rinsing the region with 0.5% chlorhexidine each individual condyloma

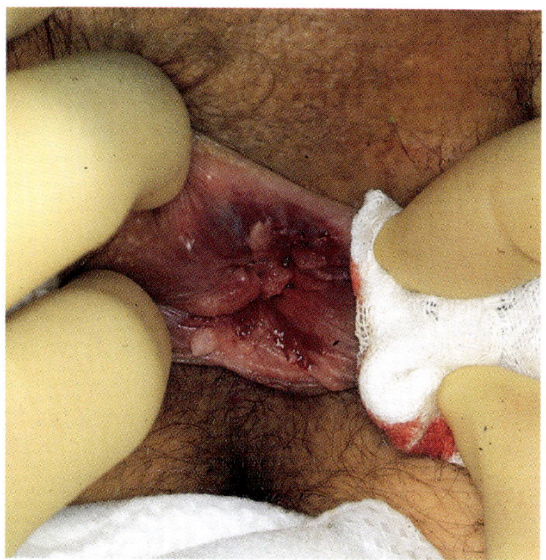

Fig. 10.9. *Anal inspection disclosing condylomata in the anal canal.*

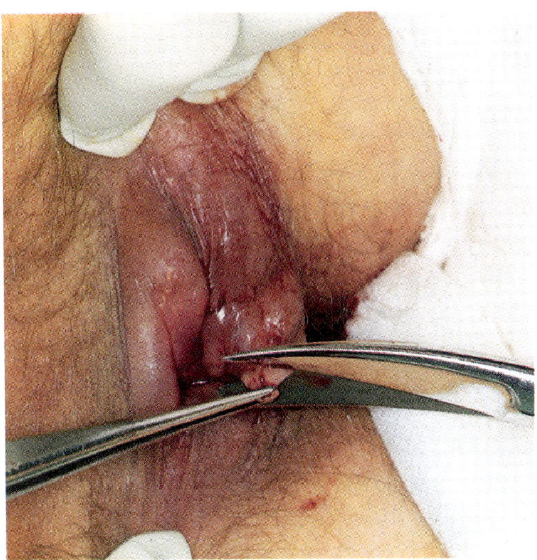

Fig. 10.11. *Simple excision of condylomata with scissors under local anesthesia.*

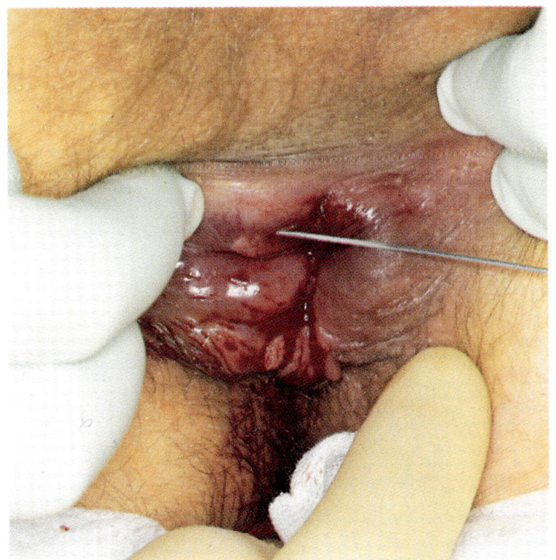

Fig. 10.10. *Local anesthesia of the condylomata from Fig. 10.9.*

is infiltrated with 1% lidocaine with noradrenaline creating small papules under each lesion. The condyloma is grasped gently with fine forceps and excised. If many condylomata are present, they can be removed at subsequent sessions. Excision causes multiple easy-bleeding small wounds. These are not sutured. The region is merely covered with gauze dressings. The patients clean the area after each motion with a shower and dry the area with a hairdryer.

Prognosis

Condylomata recur frequently after treatment. Spontaneous disappearance of condylomata can occasionally occur. On the other hand, it is not unusual to find condylomata that have been resistant to weekly podophyllin treatments that suddenly disappear. Some reach a stable number and size while others grow with time if they are not continously treated. The question of malignant transformation is as yet unanswered (see below).

Condylomata during pregnancy

Anogenital condylomata during pregnancy constitute a special problem. Condylomata are thought to be especially proliferous during pregnancy. Bacterial infection occurs easily with a consequent risk of developing sepsis. Condylomata should be treated before giving birth. Podophyllin should not be used since it carries a potential danger for systemic absorption with toxic effects on the fetus. The treatment of pregnant patients should consist of one of the local forms of destruction described earlier. Patients with intractable florid condylomata localized to the anogenital region should possibly undergo Cesarean section.

Malignant transformation

There is an undoubted connection between HPV and malignant epithelial tumors. The connection between anal condylomata and anal squamous cell carcinoma is undisputed. This verrucous carcinoma grows invasively, but metastases are very rare. HPV-DNA has been demonstrated in these tumors. The rare Buschke-Loewenstein's tumor (giant condylomata acuminata), which is also found perianally, is a form of verrucous carcinoma (a clinically malignant but histologically benign tumor). It can also contain HPV type 6. Anorectal condylomata are also associated with cervical carcinoma, and there is thus no doubt that special types of HPV virus are involved in the development of malignant epithelial tumors. From this knowledge the principle is that all tissue removed from the anal region should always be sent for pathological examination, even if one is apparently dealing with banal and condylomata. It should be realized that podophyllin-treated condylomata may display

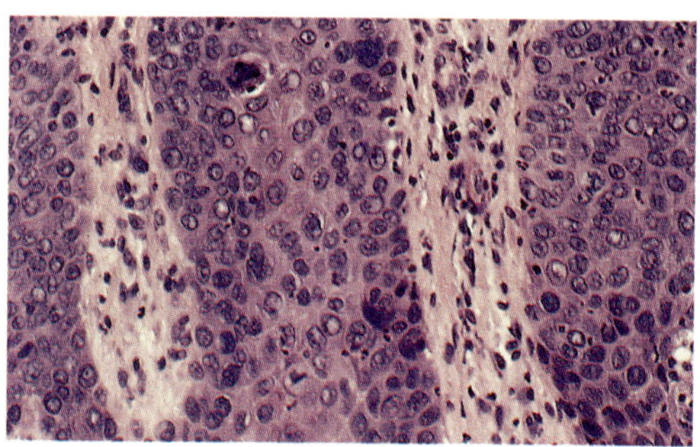

Fig. 10.12. *The effect of 4 weeks podophyllin treatment on the histological picture of condylomata – with suspicion of malignant changes.*

histological signs of malignancy (Fig. 10.12) which are in fact caused by the toxic influence of podophyllin, and which disappear after cessation of the treatment. The pathologist should, therefore, be informed that the condylomata sent for examination have been treated with podophyllin.

Prophylaxis

The use of condom should be advised as well as abstaining from oral sex with an infected partner. The patient should be informed of the nature of the disease and its methods of transmission.

SUPPLEMENTARY READING

Carr G, William DC. Anal warts in a population of gay men in New York City. Sex Transm Dis 1977; 4: 45.

Culp OS, Kaplan IW. Condylomata acuminata: two hundred cases treated with podophyllin. Ann Surg 1944; 120: 251.

Edwin DM. Condylomata acuminata: Successful treatment of four cases by hypnosis. Am J Clin Hypn 1974; 17: 73.

Jensen SL. Comparison of podophyllin application with simple surgical excision in clearance and recurrence of first episode perianal condylomata acuminata. Lancet 1985; i: 1146–9.

Montaldi DH, Giambrone JP, Courey NG, Taefi P. Podophyllin poisoning associated with the treatment of condylomata acuminata. Am J Obstet Gynecol 1974; 119: 1130.

Nel WS, Fourie ED. Immunotherapy and 5% topical 5-fluorouracil ointment in the treatment of condylomata acuminata. S Afr Med 1978; 71: 234.

Thompson JPS, Grace RH. The treatment of perianal and anal condylomata acuminata: a new operative technique. Proc R Soc Med 1978; 71: 180.

XI. Specific and nonspecific inflammation

PERIANAL FURUNCULOSIS
Definition
Acute necrotizing infection of a hair follicle in the perianal region caused by exogenously introduced staphylococci.

Epidemiology
It is found relatively infrequently in the perianal region.

Anal manifestations
A furuncle presents in the first incidence as a small follicular inflammatory nodule with redness and swelling. The lesion later becomes pustular and necrotic. The lesion usually heals up within a week after discharge of a necrotic material leaving a scar.

Treatment
The treatment of large furuncles involves incision under local anesthesia together with systemic antibiotic treatment. Before beginning treatment the pus from the furuncle should be sent for culture and determination of resistance. These studies will determine what subtype of staphylococcus is involved. Local treatment may be given with sulfa paste.

HIDRADENITIS SUPPURATIVA
Definition
A chronic recurrent inflammation of the apocrine sweat glands which may be characterized by abscess and sinus formation in the skin and subcutaneous tissue of the region.

Pathogenesis
It is believed that the disease arises in that part of the anal region that contains hair follicles, sebaceous glands, and sweat glands. The gland's duct becomes obstructed resulting in secondary inflammation which spreads to the perianal region. A secondary fistula track is formed superficially in subcutaneous tissue. Occasionally the infection is caused by clostridia and other anaerobic bacteria. The latter condition is called Fournier's gangrene and can be life-threatening.

Epidemiology
Occurs rarely. Is most often seen in young healthy patients, but also occurs in older patients with complicated diseases such as diabetes mellitus or severe arteriosclerosis.

Anal manifestations
Lesions start as multiple subcutaneous lumps that gradually grow into a plaque. Chronic fistula openings to the lower part of the anal canal occur and may extend out into the perianal region superficially to the internal anal sphincter. Recurrent pus secretion from the multiple pustula openings occurs rarely (Fig. 11.1). After a time, severe scar formation develops in the perianal region (Fig. 11.2). The perianal skin is thickened, indurated, and discolored.

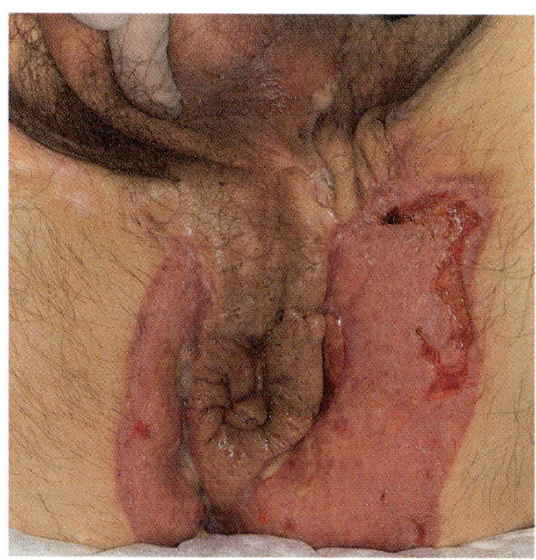

Fig. 11.1 *Recurrence of suppurative hidradenitis after surgical treatment, with pus in fistulous tracks.*

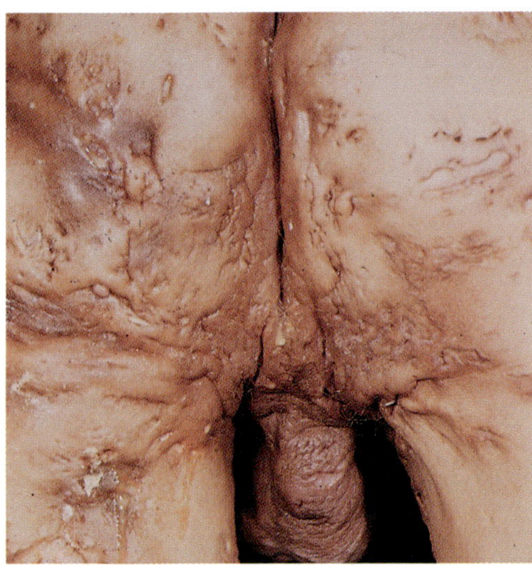

Fig. 11.2   *Chronic suppurative hidradenitis with induration of the perianal skin and multiple fistulous tracks.*

Diagnosis

The diagnosis is made from the look and location of superficial sinuses and fistulae in the
anoderm. These fistulae never penetrate the internal anal sphincter in contrast to anorectal fistulae
and Crohn fistulae, which both lie deeper.

Differential diagnoses

Crohn's disease, pilonidal cyst, furunculosis.

Treatment

Antibiotic treatment has no significant effect.
The treatment is surgical and consists of total
excision of the lesion including skin and subcutaneous tissue down to the fascia. The large
resultant wound often requires skin grafting.

ANAL TUBERCULOSIS
Definition

Infection by human or bovine tubercle bacilli
(*Mycobacterium tuberculosis* [TB]) arising in
the anal region from the bowel lumen or hema-
togenously. It is often seen together with rectal
tuberculosis.

Epidemiology

Tuberculosis is seen rarely in Denmark, but the
condition does exist; about 600 new cases of
lung tuberculosis are seen annually.

Symptoms

The condition often begins with severe pain
accompanied by bleeding and diarrhea; the
colon and rectum are involved. The patient's
general condition is affected, and there may be
weight loss and malaise. Symptoms of stricture
formation may occur.

Anal manifestations

The most common manifestation is an irregular ulcer in the anal canal or a fissure which
spreads up through the anal canal. Anal fissures
are often accompanied by incontinence. A verrucous form with irregular skin thickening can
be seen perianally. There is inguinal lymphadenopathy. Anal fistulae with or without preceding abscess formation are also described.
Anal stenosis can also occur.

Diagnosis

The diagnosis is made partly by the histological picture seen on biopsy with typical epithe-
lioid granulomata containing giant cells and
central necrosis. The diagnosis is confirmed by
the demonstration of acid-fast bacilli in the
lesion.

Treatment

Primarily medical. The main agents are isoni-
azid and rifampicin (see medical textbooks). If
abscesses or fistulae occur conservative surgical treatment by incision is indicated followed
by medical treatment.

ULCERATIVE COLITIS
Definition

Anal complications of a chronic inflammatory
disease of the mucosa of the colon and rectum of unknown but probably multifactorial
etiology.

Epidemiology

The disease is seen more commonly in females
than in men (a ratio of 1.3:1). The incidence is
about 7–8 per 100,000 population. Occurs in
both young and older patients. The incidence of
accompanying anal lesions is no different from
the incidence in the normal population.

Pathological anatomy

The disease is not a distinct histopathological entity. On rectoscopy the mucosa is found to be hyperemic with punctate bleeding. Ulcerations occur later in the course of the disease. The mucosa may be covered with mucus, blood, and pus. Microscopic microabscesses in the mucosa and vasculitis are found. Changes are confined to the mucosa and submucosa.

Symptomatology

It depends on the anal lesion involved. The specific anal symptom is usually accompanied by bloody diarrhea originating from the rectum.

Anal manifestations

Hemorrhoids (20%), occasionally fissures (6%), and anal abscesses or fistulae (2–6%).

Differential diagnoses

Common anal diseases, Crohn's disease, anal tuberculosis.

Treatment

Symptoms from the anal lesions are worst during the active phase of the disease, with copious diarrhea. Anal symptoms usually disappear on medical treatment. Anal abscesses arising should be incised. There is nothing to indicate that one should be surgically conservative in treating the anal lesions of ulcerative colitis (in contrast to those occurring with Crohn's disease).

ANAL CROHN'S DISEASE

Definition

Perianal or anal lesions in patients with Crohn's disease. Crohn's disease can be defined as an inflammatory intestinal disease of unknown origin which involves all layers of the bowel. The disease may present as diffuse lesions from the lips to the anus. It is most often seen in the terminal ileum, colon, and is extremely rare in the esophagus and stomach. It is unclear whether the anal lesions of Crohn's disease should be considered as a complication or a local manifestation of the primary disease.

Epidemiology

Crohn's disease is most commonly seen in young persons. The average age is about 30 years, and about 25% of the patients are under 20 when the diagnosis is first made. The yearly incidence in Denmark is 4–5 per 100,000 of the population. The incidence of Crohn's involvement of the anal region varies from 8 to 95% depending on the diagnostic criteria, whether the data is collected retro- or prospectively, the severity of the disease, and the definition of the anal lesions. If a study includes, for example, cases of ulcerative colitis the incidence of anal lesions will be lower than it should be, since ulcerative colitis is not thought to be accompanied by a higher incidence of anal lesions. Similarly, the incidence is lower in those studies that only include fissures and anal fistulae as Crohn lesions compared with studies also including skin tags as Crohn lesions. About 75% of patients with colonic Crohn's disease have anal lesions in contrast to only 25% of those with small bowel Crohn.

Symptomatology

It is characteristic for Crohn lesions that they are relatively asymptomatic. Pain is usually the result of secondary infection of the lesions. Edematous skin tags and/or external hemorrhoids (Fig 11.3) should arouse suspicion of Crohn's disease.

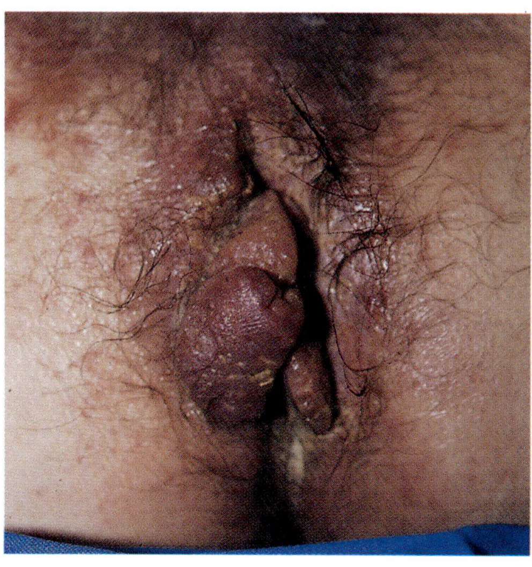

Fig. 11.3   *Edematous skin tags in a patient with Crohn's disease.*

Diagnosis

If the patient has already been diagnosed as suffering from Crohn's disease based on the usual diagnostic criteria, most anal lesions are classified as anal Crohn's disease. It is, however, doubtful whether all such lesions are caused by Crohn's disease. On the other hand, it is clear that the treatment of the anal lesions occurring in a Crohn patient should be conservative whether they are caused by the disease or have merely arisen coincidentally.

The examining doctor should bear Crohn's disease in mind when encountering a relatively asymptomatic and painless anal fissure (Fig. 11.4) or ulcerations. These will often be deep, multiple, and eccentric. The anal fissures have cyanotic edges and frequently occur laterally as opposed to common anal fissures which are most often seen posteriorly. All forms of anal stenoses, anal abscesses, anorectal or rectovaginal fistulae which are resistant to treatment and run a chronic course should suggest the presence of underlying Crohn's disease.

Diagnosis is made from a thorough rectoscopy with biopsy of any mucosal changes, and X-ray examination of the colon and the small intestine.

Skin tags are often seen in Crohn's disease. They are often bigger and thicker than the skin tags of patients without Crohn's disease.

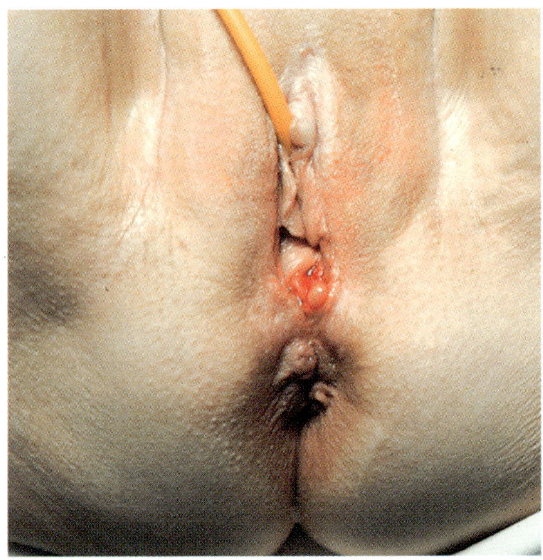

Fig. 11.5   *Chronic anal fistula in a patient with Crohn's disease.*

Fistulae in Crohn's disease are often multiple and may be situated far from the anal region (Fig. 11.5) for example in the scrotum, labia or thighs. The fistulae are usually painless and indurated (Fig. 11.6). They are less often preceded by abscess formation than the commonly occurring fistulae.

Patients with Crohn's disease rarely have internal hemorrhoids, but they often have edematous skin tags which are sometimes incorrectly diagnosed as hemorrhoids, resulting in

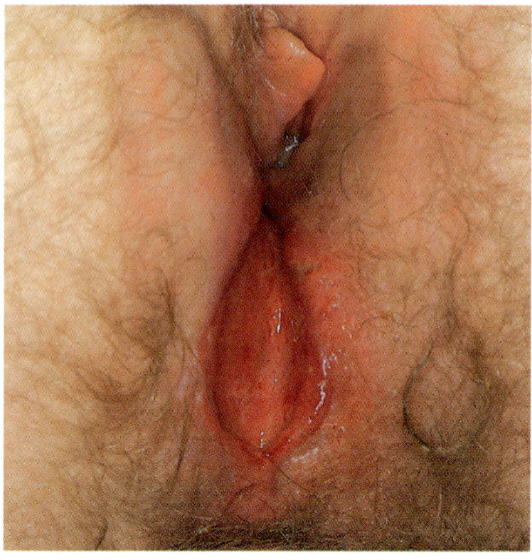

Fig. 11.4   *Asymptomatic anal fissure in a patient with Crohn's disease. A typical Crohn skin tag is also seen.*

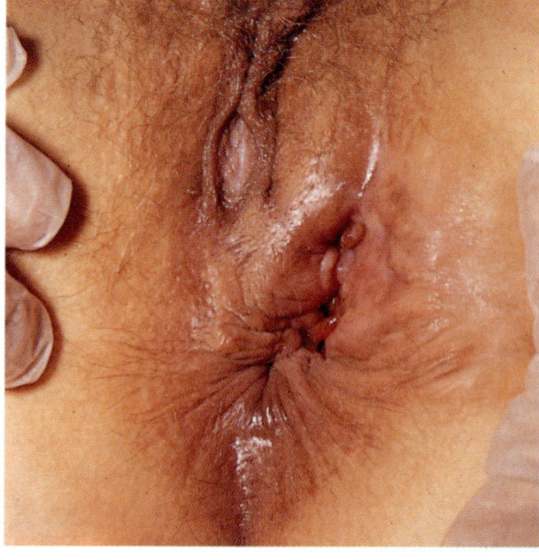

Fig. 11.6   *Chronic anal fistula with induration of the surroundings in a patient with Crohn's disease.*

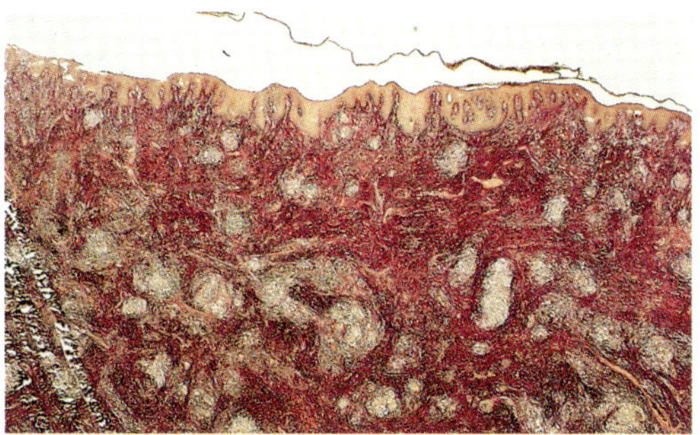

Fig. 11.7   *Anal biopsy demonstrating typical granulomas in a patient with Crohn's disease.*

inappropriate treatment. All lesions of the anus and perianal region must be biopsied. The presence of sarcoid granulomata (Fig. 11.7), are found in 40–60% of Crohn's disease.

The number of granuloma vary from a few to many in the dermis, subcutaneous tissue, and/or fistula track, and will sometimes depend on the size of the biopsy. Small biopsies are often useless. The diagnosis should be considered when anal lesions are accompanied by intestinal symptoms even if granulomata are absent.

Differential diagnoses

Sarcoidosis and tuberculosis as well as anal lesions of other etiologies. The differential diagnosis between Boeck's sarcoidosis, tuberculosis, and Crohn's disease cannot be made from histological criteria alone. The presence of caseous necrosis should naturally arouse the suspicion of tuberculosis. If it is absent, the diagnosis should then be made on culture.

One should suspect sexual diseases of the anal region as a differential diagnosis to Crohn's disease when confronting anal lesions that are unaccompanied by bowel symptoms. The diagnosis of the different sexual disorders is made as previously described (see chapter 9). In patients in poor general health one should consider whether the anal lesions are an expression of leukemia. Bilharzia of the anal region is seen in the Middle East and in tropical regions.

Complications

The following complications of Crohn's disease of the anal region occur: perianal abscess for-

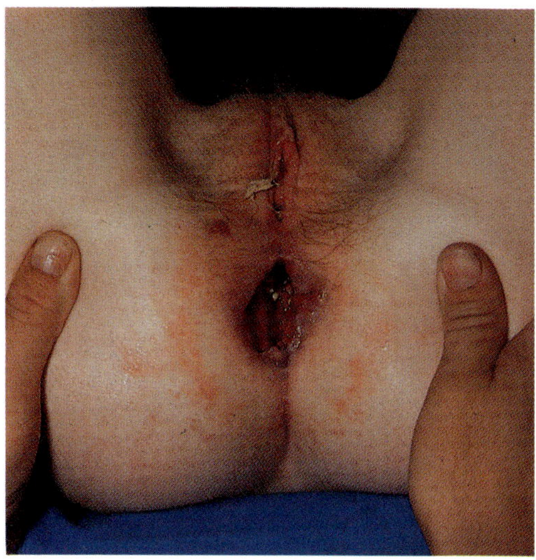

Fig. 11.8   *Anal stenosis occurring after surgery for piles in a patient with Crohn's disease.*

mation, anal stenosis (Fig. 11.8), incontinence, and carcinoma.

Abscess formation is probably caused by infection of the anal glands and arises from retention of pus. Anal stenosis often results from healing of deep anal fissures by fibrosis. In such cases, symptoms of stenosis are often mild. Serious stenosis symptoms are as a rule seen only as complications of unnecessary surgical treatment of, for example, hemorrhoids and anal fistulae. Surgery of the anal lesion in Crohn's disease can result in incontinence. The development of carcinoma in the anal lesions of Crohn's disease occurs very rarely.

Treatment

The guideline in the treatment of anal lesions in Crohn's disease is to be conservative and avoid aggressive surgery. This also applies to anal abscesses which should be opened with care. If spontaneous perforation has occurred no further treatment should be undertaken. We believe the only indications for surgical treatment are incision of acute absceses and careful dilatation of symptomatic anal stenoses. Extensive surgical procedures in anal fistulae may result in non-healing fistulae and anal incontinence, or worsening of an already existing incontinence.

The conservative treatment consists of rest or systemic treatment with steroids, azathioprin or antibiotics such as metronidazole.

Prognosis

About 10% of patients will require proctectomy during the course of the disease, 40% will resolve spontaneously with time, and about 30% will become asymptomatic.

ANORECTAL BILHARZIA

Definition

Anorectal manifestations of infestations with *Schistosoma* eggs. The different types of schistosomiasis occur according to the morphology of the eggs: *Schistosoma haematobium, S. mansoni* or *S. japonicum.*

Mode of infestation and incidence

Infested patients excrete *Schistosoma* eggs in the urine or feces. The eggs hatch in fresh water, and the larvae infest specific species of snails in which many thousands of new larvae develop. These larvae infest humans coming into contact with infested fresh water by penetrating the skin. About 300 million people world-wide are infested with one of the three species of *Schistosoma*. The disease is especially prevalent in the developing countries of Africa, South America, and the East.

Symptomatology

S. mansonia and *S. japonicum* give gastrointestinal symptoms, while *S. haematobium* causes symptoms in the urinary tract. The patients are often asymptomatic for many years. Patients infected with *S. mansoni* or *S. japonicum* often complain of abdominal colic accompanied by bloody diarrhea.

Anorectal manifestations

Perianal bilharziasis

Perianal bilharziasis is an extremely rare manifestation. It appears as an irregular indolent tumor that grows progressively. Diagnosis is made histologically by demonstrating the intradermal presence of *Schistosoma* eggs together with intracutanous foreign body granulomata. This is a result of migration of the eggs to the skin.

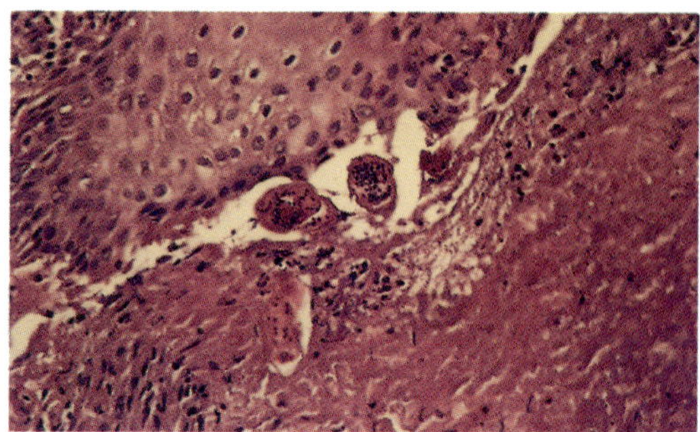

Fig. 11.9 *Biopsy from the perianal region of a patient with Bilharziasis.*

Rectal bilharziasis

The rectal changes of bilharziasis may resemble diffuse polyposis. Pseudopolyposis, mucosal hyperemia with punctate ulcers and rectal varices in the form of dilated veins may be found.

Diagnosis

Diagnosis is made histologically from rectal biopsies or biopsies from the perianal skin (Fig 11.9). Demonstration of *Schistosoma* eggs is mandatory for the diagnosis. The eggs can be found in the feces. Diagnosis can also be made from skin and complement fixation tests.

Treatment

Niridazol (Ambilhar® Ciba-Geigy) (25 mg/kg for 7 days) is an effective drug in the treatment of bilharziasis. The drug should not be used in the presence of liver damage resulting from *S. mansoni* infestations. Trivalent antimony compounds should be used instead. These are always used for *S. japonicum* infestations. The anal changes usually disappear with this treatment.

Prophylaxis

Elimination of snails. Avoiding bathing in fresh water in infested areas.

ANORECTAL AMEBIASIS

Definition

Infection of the rectum or perianal skin area caused by the protozoan *Entamoeba histolytica*. The protozoan occurs in two forms, a trophozoite form and a cyst form.

Pathogenesis

Infection arises from oral intake of an acid-resistant cyst which multiplies in the bowel and develops into an intestinal amebiasis which may be asymptomatic if the bowel wall is not invaded. If the bowel wall is invaded a series of small ulcers is seen in the mucosa of the rectum and colon. Spreading can occur from the rectum to the perianal skin surface, and fistulae may arise with surrounding cutaneous amebiasis. Metastatic spread can occur to liver, lungs, and brain as well as to other organs.

Incidence

Invasive amebic dysentery is seen only in warm climates while non-invasive amebae occur universally.

Symptomatology

When the cyst invades the bowel wall either acute or chronic amebic dysentery will result. In the acute phase the patients have frequent thin, slimy, bloody motions without pus. Tenesmus is common. The chronic form presents with recurrent acute attacks with asymptomatic intervals.

Perianal amebiasis is extremely rare. It presents as an anal fistula with cutaneous amebiasis or simply as cutaneous amebiasis alone. The cutaneous amebiasis presents as ulceration of variable and irregular extent, but with sharp, distinct edges. The ulcer is often covered by purulent secretion.

Diagnosis

The diagnosis is made at rectoscopy. The characteristic ulcerations are often found in the rectum. These are yellowish with sharp edges and surrounded by a hemorrhagic halo. There is central crater formation containing necrotic material. The changes are spread irregularly in the rectum, and a diffuse yellowish discharge is found arising from the ulcers. It is important for the diagnosis to demonstrate the trophozoites in the acute phase. These can be found in the feces or in the discharge from the ulcers. An easy method of "catching" the trophozoites is to suck mucus from the rectal ulcerations up into

the meshes of a small piece of Spongostan®, fix it in Bouin's fluid, section the block in a microtome and demonstrate the protozoa with ordinary histological techniques. Trophozoites as well as cysts can be seen in the biopsy material from ulcerations in the rectum.

With cutaneous amebiasis the diagnosis is made by showing the protozoa in biopsy material from the perianal skin lesions.

Treatment

The standard treatment is oral metronidazole 800 mg t.d.s. for 5 days. In severe cases this treatment is not sufficient and it should be supplemented with tetracyclin and emetin (see internal medicine textbooks). Large doses of metronidazole (2,400 mg daily for 2 days) is required for metastatic spread of the disease, for example, to the liver.

Follow-up

The feces are examined for cysts from 4–6 weeks after completion of the treatment in cases of intestinal amebiasis. Ten negative examinations are required before the patient can be regarded as cured. One should be aware that cyst excretion might be intermittent.

SUPPLEMENTARY READING

Alexander-Williams J. Perianal Crohn's disease. World J Surg 1980; 4: 203.

Broadwater JR, Bryant RL, Petrino RA et al. Advanced hidradenitis suppurativa: review of surgical treatment in 23 patients. Am J Surg 1982; 144: 668.

Culp CE. Chronic hidradenitis suppurative of the anal canal – a surgical skin disease. Dis Colon Rectum 1983; 26: 669.

Hobbis JH, Shofield PF. Management of perianal Crohn's disease. J R Soc Med 1982; 75: 414.

Marks CG, Ritchie JK, Lockhart-Mummery HE. Anal fistulas in Crohn's disease. Br J Surg 1981; 68: 525.

Thorntin JP, Abcarian H. Surgical treatment of perianal and perineal hidradenitis suppurativa. Dis Colon Rectum 1978; 21: 573.

Williams DR, Coller JA, Corman ML et al. Anal complications in Crohn's disease. Dis Colon Rectum 1981; 24: 22.

Wolff BG, Culp CE, Beart RW Jr et al. Anorectal Crohn's disease: a long-term prospective. Dis Colon Rectum 1985; 28: 709.

XII. Benign and premalignant tumors

Many different kinds of benign tumors occur in the anal region, especially perianally. Benign tumors arising from the region's connective tissues as well as those of the perianal apocrine sweat glands occur very rarely. Sporadic cases of leiomyoma, neurofibroma, and squamous cell myoblastoma are described in the literature. Leiomyomata arise from the internal anal sphincter (smooth muscle), while myoblastomata occur in striated muscle.

Only those tumor types that are likely to be encountered even in a coloproctology service will be discussed here, since knowledge of their correct treatment is important.

DYSPLASIA AND CARCINOMA *IN SITU*

Definition

Lesions of the perianal skin or anoderm with all the cytological signs of malignancy, but with no spread of the disease, that is to say, the basal membrane remains intact. The epithelium is dedifferentiated with distorted layering and loss of cellular polarity. The changes may be graded into mild, moderate or severe dysplasia as well as carcinoma *in situ*. Carcinoma *in situ* is also called intraepithelial carcinoma.

Types

Carcinoma *in situ*, Bowen's disease, Queyrat's erythroplasia, and Paget's disease. Queyrat's erythroplasia is found most often on the penis, but the authors have seen cases in the perianal skin.

Epidemiology

Varying grades of dysplasia and/or carcinoma *in situ* changes are often found around the edges of an invasive squamous cell carcinoma of the anal canal. These changes can also be seen widely distributed in the anal canal surrounded by normal tissue remote from the zone around an invasive carcinoma. These changes have not been demonstrated with adenocarcinoma in the region.

Severe dysplasia and carcinoma *in situ* changes are also seen in the absence of invasive squamous cell carcinoma. In Brazil, dysplasia and carcinoma *in situ* changes have been reported as a routine finding on microscopic examination of anal tissue removed in connection with, for example, hemorrhoidectomy in about 2% of the cases. These changes are seen only in 0.2% of the cases in the USA. Spontaneous dysplasia and carcinoma *in situ* changes are described in the transitional epithelium without accompanying invasive cancer. In the routine examination of tissue removed from the anal canal in non-malignant conditions, incidence of all grades of dysplasia is given as 0.3 per 100,000 of the population.

Symptoms

Patients are usually asymptomatic.

Diagnosis

The only way of making the diagnosis is by biopsy of the suspicious area. A colposcope may be used, especially in the lower end of the anal canal. All tissue, even that removed in surgery of common anal conditions, should be sent for histological examination, even though the incidence of dysplasia and carcinoma *in situ* changes in the absence of invasive cancer is small.

Treatment

Local excision. With diffuse perianal changes laser surgical excision can be considered with skin flaps to obtain wound cover. In treating these conditions it should be remembered that the lesions may occur diffusely with normal intermediate tissue.

Results

Patients with carcinoma *in situ* changes should be monitored for life, since the recurrence rate is high after local excision of this condition. On follow-up the whole anal canal should be examined thoroughly, including the transitional zone, that is to say, at least 2–3 cm proximal to the dentate line.

LEUKOPLAKIA

Definition

A term used for a chronic, whitish superficial lesion of the skin and mucosa of the anal region (Fig. 12.1). Histologically, hyperkeratosis and irregular acanthosis together with atypical cells and lymphocyte infiltration of the underlying connective tissue is found.

Epidemiology

This occurs relatively frequently, but the incidence is unknown, since not all surgeons routinely send all tissue that is removed from apparently benign lesions of the anal region for microscopic examination.

Etiology

Unknown. It is thought that the condition is an expression of a prolonged and persistent irritation, for example, a non-specific chronic inflammatory condition. Pruritus ani, anal fissures, and hemorrhoids are thought to predispose to the condition.

Symptomatology

No specific symptoms. Any symptom will depend on the primary disorder. Symptoms in the form of itching, burning or stinging may occur, especially after defecation.

Treatment

Surgical excision. If the diagnosis of leukoplakia is made incidentally on routine microscopy of removed tissue, and such tissue has been removed to within healthy margins, nothing further is done. If this is not the case, the patient should be reoperated. Patients should be followed regularly, that is, once a year for 5 years, since the condition often recurs.

Prognosis

Good. There is no basis for regarding leukoplakia of the anal region as a premalignant condition in contrast to leukoplakia in the buccal cavity.

KERATOACANTHOMA

Definition

An elevated dome-shaped lesion consisting of squamous epithelium with massive keratin formation which fills a central crater. The tumor resembles a malignant squamous cell carcinoma both clinically and histologically, but it behaves like a benign tumor and grows much more rapidly than a squamous cell carcinoma.

Pathological anatomy

Histologically keratoacanthoma can be divided into three types: solitary, multiple, and eruptive. Only the solitary form has been described in the anal region. The tumor occurs all over the body, but most frequently on areas exposed to sun.

Microscopically the tumor is usually not distinguishable from an ordinary squamous cell carcinoma (see Fig. 13.4). It is a dome-shaped, invaginating epidermal tumor with the classical central keratin plug (Fig. 12.2).

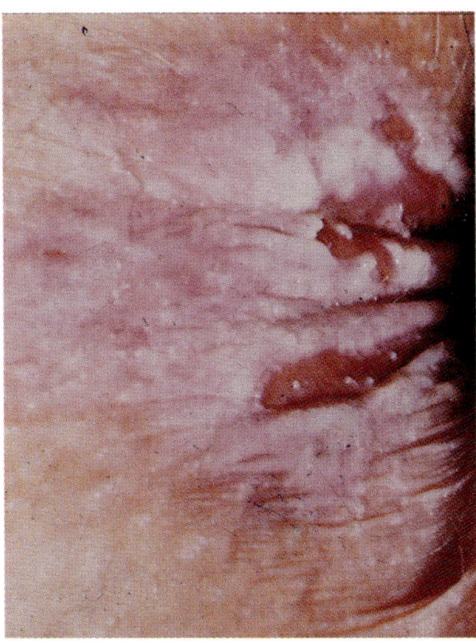

Fig. 12.1 *Macroscopic appearance of leukoplakia of the perianal region.*

Fig. 12.2 *Histological appearance of a keratoacanthoma of the anus.*

Epidemiology

The tumor occurs frequently on areas of the body exposed to the sun, especially the face and the forearms. The greatest incidence is amongst women in the sixth to seventh decade. The anal site is seen very rarely. Only five cases have been described in the literature, three of which were from Denmark. This may be due to the fact that the tumors are classified as squamous cell carcinomas with an actual incidence much greater than that reflected in the literature.

There is evidence to suggest that patients with keratoacanthomas have a high incidence of visceral tumors, especially of the colon.

Etiology

Unknown.

Symptomatology

The lesion presents as an ulcerating, dome-shaped lesion of the anal canal or anal edge without induration of the region. They are occasionally accompanied by bleeding or pruritus ani.

Diagnosis

The most important aid in the diagnosis of keratoacanthoma is the duration of the tumor. If it has been present for more than 3–6 months and if the histological picture resembles squamous cell carcinoma the tumor should be regarded as a carcinoma. Keratoacanthomas resolve spontaneously in the course of 8–24 weeks. The tumor is reddish with central crater formation. It is important to take a representative biopsy.

Treatment

Local excision or curettage.

Prognosis

Good. Only 12% of the tumors recur after an apparently radical treatment. Of these about 1–2% have transformed to squamous cell carcinomas.

PAGET'S DISEASE OF THE PERIANAL REGION
Definition

Paget's perianal disease is histologically identical to Paget's disease of the breast. The lesion is defined as an intraepidermal neoplasm. Characteristic for the tumor are the large, pale, irregular Paget cells.

Pathological anatomy

Microscopically the tumor's appearance and size is very variable. The lesion may resemble a common eczema. There is no actual elevation of the skin (Fig. 12.3) but induration is present. Alternatively, the tumor may resemble a hemorrhoid. The tumor can be benign or malignant and is accompanied in about 60% of cases by a malignant visceral tumor or an underlying adenocarcinoma arising from the apocrine glands of the skin.

Microscopically the epidermis is infiltrated with the irregular, large, pale Paget cells which lie indisposed amongst apparently normal squamous epithelial cells (Fig. 12.4). If the Paget

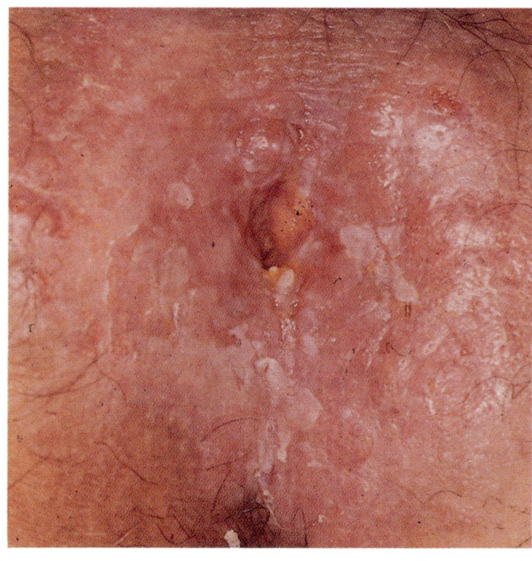

Fig. 12.3 *Paget's disease of the perianal skin.*

cells infiltrate apocrine structures with an intact basal membrane, this is regarded as *in situ* Paget's disease. On the other hand, if infiltration of the apocrine structures of the skin is shown with invasion through the basal membranes of the surrounding tissue, this is called invasive Paget's disease. Paget's disease thus occurs in three forms: non-invasive, *in situ*, and invasive.

Epidemiology

Paget's perianal disease is thought to be a rare condition. Until now only about 100 cases have been described in the literature. The true incidence is not known. It is seen at all ages, but most commonly amongst women in the sixth to seventh decade. About 60% of the tumors are non-invasive, 20% *in situ*, and 20% invasive malignant tumors. About 40% of the patients also have an accompanied malignant tumor in the gastrointestinal tract (rectum, colon).

Symptomatology

The most common symptom is severe itching which is seen in about 75% of the patients. The condition is often regarded as eczema (Fig. 12.3). Other symptoms are an impression of a mass (40%), perianal soiling (40%), and change in bowel habit (32%). The symptoms have usually been present for a long time before the correct diagnosis is made.

Diagnosis

The diagnosis is made on histological examination (Fig. 12.4). The Paget cells are positive on aldehyde-fuchsin staining (pH 1.7) in contrast

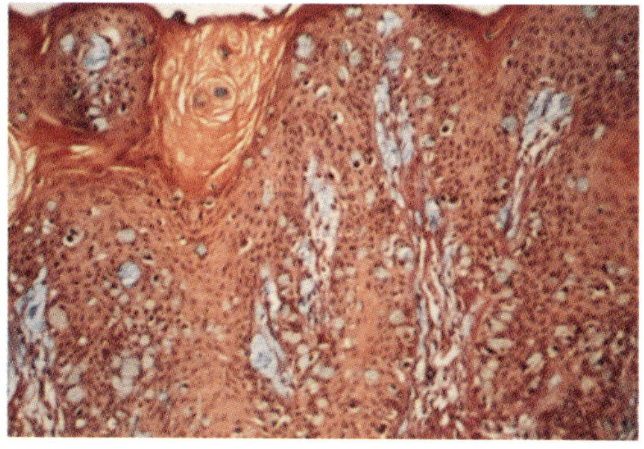

Fig. 12.4   *Histological picture of Paget's perianal disease. The Paget cells appear as nests or single cells in the epidermis.*

to the cells of malignant melanomas, Bowen's disease, leukoplakia, and squamous cell carcinoma. When investigating a patient for Paget's perianal disease it is important to examine the gastrointestinal tract thoroughly for malignant tumors.

Differential diagnoses

The differential diagnoses and the most commonly occurring misdiagnoses are: perianal eczema, chronic lichen simplex, condylomata acuminata, psoriasis, hemorrhoids, and hyperkeratosis.

Treatment

Paget's disease without invasive features as with an *in situ* form can be treated by wide local excision with subsequent plastic surgical skin cover, possibly with a temporary colostomy. Laser excision has given promising results. The treatment for invasive Paget's disease is abdominoperineal excision of the rectum. If there are accompanying malignant tumors these must be treated according to the usual guidelines. Radiation or chemotherapy have no place in the treatment of invasive Paget's disease.

Prognosis

Recurrence is common after less radical procedures such as local excision and curettage. Some of the recurrences are invasive. The recurrence rate is about 50% within the first 5 years (Fig. 12.5). The survival of these patients is reduced. In a study, the 5-year survival rate was 54% for patients with Paget's perianal disease in contrast to 84% for the normal population.

BOWEN'S PERIANAL DISEASE
Definition

An intraepidermal squamous cell carcinoma also regarded as carcinoma *in situ*. It occurs in both sun-exposed skin and skin not exposed to the sun, and can be accompanied by visceral cancers.

Pathological anatomy

The typical Bowen element is seen macroscopically as a slow-growing, erythematous, desquamating lesion which is sharply, but

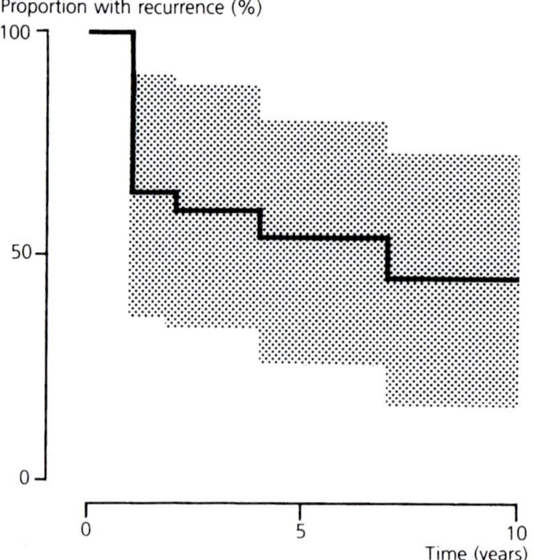

Fig. 12.5   *Recurrence rates after initial cure for Paget's perianal disease.*

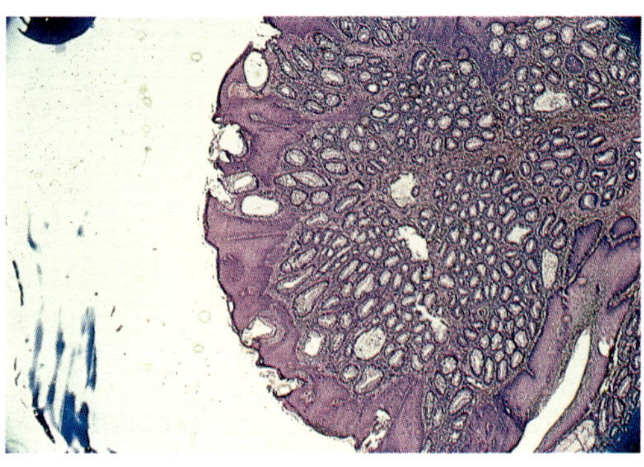

Fig. 12.7   *Histological appearance of Bowen's disease of the perianal region.*

irregularly delineated from the surrounding skin (Fig. 12.6). It may occur as solitary elements in the perianal region or diffusely around the whole anal circumference. The lesions recur in about 30% of the cases after removal. The recurrence will be invasively infiltrating in about 1–2% of the cases.

Microscopically (Fig. 12.7) acanthosis of the epidermis is found, and the cornified layer is thickened and consists mainly of parakeratotic cells. The cells lie irregularly in the epidermis.

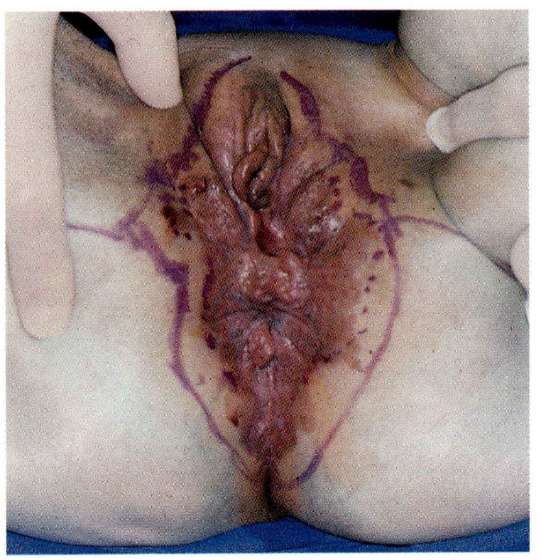

Fig. 12.6   *Macroscopic appearance of perianal Bowen's disease.*

Large round dyskeratotic cells with a homogenous eosinophilic cytoplasm and pyknotic nuclei are found.

Etiology

It is thought that there is a connection between the development of visceral malignant tumors and Bowen's disease. We have, however, been unable to prove this in a Danish study of 48 patients with perianal Bowen's disease. Some of the tumors are caused by ingestion of arsenic.

Epidemiology

Bowen's disease of the perianal region is a rare premalignant lesion. Women are affected more commonly than men (2:1). The condition is seen most frequently in patients of 40–50 years of age.

Symptomatology

The patients often complain of pruritus ani or of a desquamating mass or induration in the perianal region.

Diagnosis

The diagnosis is made histologically from good, representative biopsies. As there probably is no significant relationship between Bowen's disease of the perianal region with accompanying visceral tumors, there is no need for extensive examinations in search of such tumors.

Differential diagnoses

Bowenoid papillomatosis (multiple, lightly pigmented perianal papillae that are histologically identical with Bowen's disease), Paget's perianal disease, squamous cell carcinoma, basal cell carcinoma, leukoplakia, and arsenic keratosis.

Treatment

The treatment consists of local surgical excision. In large lesions, for example, those including the labia in women, the wound is covered by rotation of skin flaps together with a temporary colostomy. Usually the lesions are small and allow conservative treatment in the form of cryotherapy, electrocoagulation, laser treatment or 5-fluorouracil. Larger lesions as well as smaller non-invasive recurrences can be treated successfully by large laser surgical excisions.

Prognosis

The prognosis is good, but long-term follow-up is recommended, since recurrence may occur 13 to 15 years after primary radical surgery. Only 6% will develop an invasive squamous cell cancer requiring radical surgery.

ANAL HEMANGIOMA
Definition

A benign vascular tumor which most often is congenital, but can occur later, and should be regarded as a true tumor. The tumors may be divided into capillary or cavernous hemangioma according to the size of the vessels involved.

Pathological anatomy

When the tumor occurs in the rectum or anal canal it is most often a solitary, diffusely infiltrating, cavernous hemangioma. It starts most frequently in the mucosa and gradually narrows the anus (Fig. 12.8). Microscopically a thick-walled vascular space covered by endothelium is found. The vessel walls are composed of smooth muscle and connective tissue.

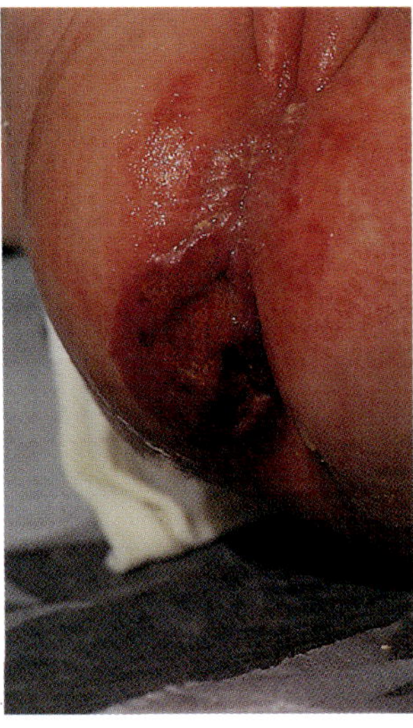

Fig. 12.8
Hemangioma of the anus.

Epidemiology

These tumors are extremely rare, and the individual proctologist is unlikely to see a case during his career.

Symptomatology

Bleeding, which may be massive, prolapse symptoms, or sometimes symptoms of stenosis.

Treatment

Local excision when this is possible, otherwise abdominoperineal resection of the rectum if severe uncontrollable bleeding occurs.

SUPPLEMENTARY READING

Beck DE, Fazio VW. Perianal Paget's disease. Dis Colon Rectum 1987; 30: 263.

Jensen SL, Sjølin KE. Kerathoacanthoma of the anus: report of three cases. Dis Colon Rectum 1985; 28: 743.

LaVoo JW. Bowenoid papulosis. Dis Colon Rectum 1987; 30: 62.

Quan SHQ. Anal and perianal tumors. Surg Clin North Am 1978; 58: 591.

Scoma JA, Levy E. Bowen's disease of the anus. Dis Colon Rectum 1975; 18: 137.

Strauss RJ, Fazio VW. Bowen's disease of the anal and perianal area. Am J Surg 1979; 137: 231.

William SL, Rodgers LW, Quan SHQ. Perianal Paget's disease: report of 7 cases. Dis Colon Rectum 1976; 19: 30.

XIII. Malignant tumors

Malignant tumors of the anal region occur about 20–30 times less frequently than colorectal tumors. In a retrospective study in Denmark covering the period from 1943 to 1973, an incidence of only 0.2 per 100,000 of the population was registered (an investigation of historical interest). More recent prospective studies reveal an incidence of 0.7 per 100,000 of the population.

Despite the early symptoms and superficial location of these tumors it is unfortunate that both the patient and the doctor often misdiagnose them and regard the symptoms as arising from one of the more common benign disorders of the region. The prognosis is therefore still bad.

Because of the tumors' rarity most studies reported in the literature are small and many questions concerning their treatment and prognosis are controversial.

The malignant tumors of the anal region include a whole series of different histological types, namely basal cell carcinoma, squamous cell carcinoma, adenocarcinoma, malignant melanoma, as well as the extremely rare malignant tumors arising from connective tissue, fat, and muscle.

It is important to classify malignant tumors of this region into perianal tumors and tumors of the anal canal, since they are biologically and prognostically different.

The incidence of the tumors in the two sites depends on how the perianal region is defined. If the anal verge is regarded as the border between the anal canal and the perianal region, about 15% of the tumors will be classified as perianal. If, on the other hand, the border is defined as the dentate line, about 30% will be classified as perianal tumors.

The authors have always opted to consider the anal verge as the border between the two regions of the anus.

Although rare, anal manifestations of malignant hematological diseases are also seen. The treatment of these anal complications will be discussed only briefly.

BASAL CELL CARCINOMA
Definition
A slow-growing, locally invasive tumor that metastasizes very rarely and is, therefore, often regarded as benign in practice.

Pathological anatomy
Macroscopically the typical tumor presents in the anal canal as a lump with central necrosis and elevated, rounded edges. The tumor is seen most frequently lateral to the anal opening as a solitary lump, but on extremely rare occasions a multiple form involving the whole anal circumference is seen.

Microscopically, the tumor has a characteristic appearance (Fig. 13.1). Lobules or strands containing atypical basal cells are seen in the chorium. These are arranged peripherally in a disorderly palisade-like fashion. The tumor cells have large, oval basophilic nuclei surrounded by sparse cystoplasm.

Epidemiology
The tumors occur on the sun-exposed regions of the body (especially on the face), but are also seen in the anal region. Basal cell carcinomas in the perianal region, however, are rare. The

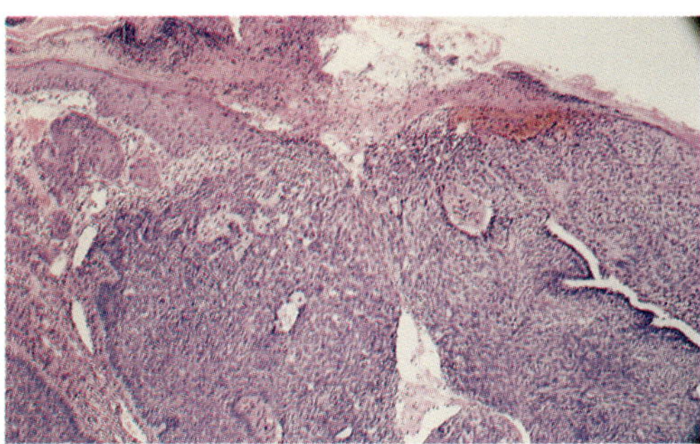

Fig. 13.1. *Histological picture of a basal cell carcinoma of the perianal region.*

tumors are seen with equal frequency in men and women, usually in the sixth to seventh decade.

Symptomatology

The most common symptom is a feeling of a mass in the anal region with ulceration, tenderness, and bleeding. There may be associated lymphadenopathy in the inguinal regions.

Diagnosis

The diagnosis is made from the lesions' typical appearance and from histological examination of a representative biopsy.

Differential diagnoses

Keratoacanthoma, squamous cell carcinoma, Bowen's perianal disease, Paget's perianal disease.

Treatment

Radiation therapy

The tumor is radiosensitive, and this treatment is thought to be appropriate for large tumors lying peripherally in the perianal region in elderly patients. Recurrences after radiation therapy should be surgically excised.

Surgical excision

Local excision with primary closure of the wound is the standard treatment for this kind of tumor and can be done under local anesthesia without problems. Hospitalization during treatment is only necessary if the lesion is large. Skin transplants might be necessary to achieve skin cover.

When surgical removal of these tumors is undertaken, one should be aware of the fact that their extent is often underestimated, and the excision should be deep and wide. It is important to make sure that the lesion has been fully removed by identifying sound tissue on histological examination. The authors have seen a case where spreading to the inguinal lymph nodes had occurred. Block resection of the lymph nodes cured the patient.

Prognosis

The prognosis is good. Mortality is not increased. The recurrence rate is low, especially after surgical excision.

SQUAMOUS CELL CARCINOMA

Definition

Biologically aggressive tumors that apparently arise in the squamous cell layer of the anal region without any preceding features. According to the cell types, squamous cell carcinomas are divided pathoanatomically into ordinary, basaloid, and mucoepidermoid tumors. This classification is not thought to have any clinical value.

Pathological anatomy

Ordinary squamous cell carcinomas

Ordinary squamous cell carcinomas (Figs. 13.2 and 13.3) may be cornified or uncornified according to their location in the anal region. Perianal tumors are most often cornified and highly differentiated in contrast to tumors of the anal canal which are differentiated and

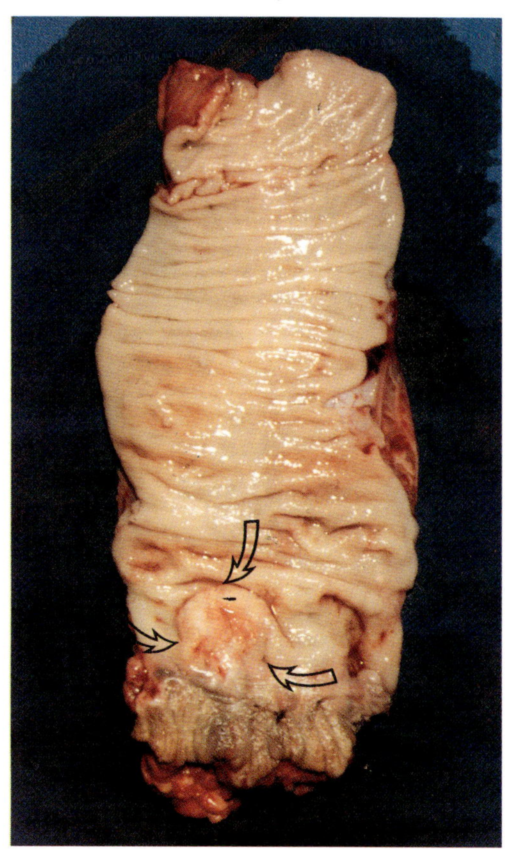

Fig. 13.2. *Squamous cell carcinoma of the anal canal (surrounded by arrows).*

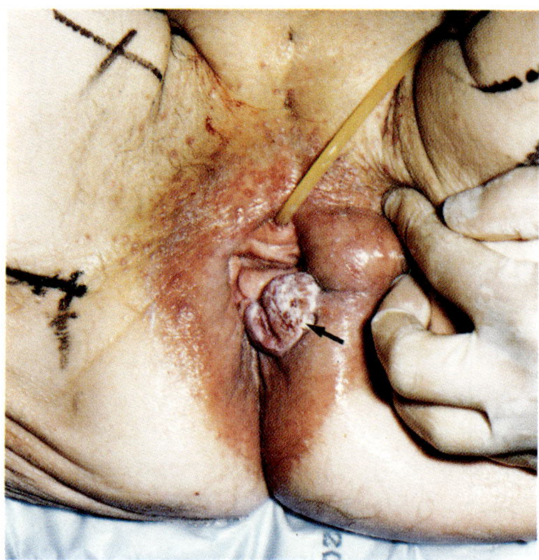

Fig. 13.3. *Macroscopical appearence of a squamous cell carcinoma of the perianal region.*

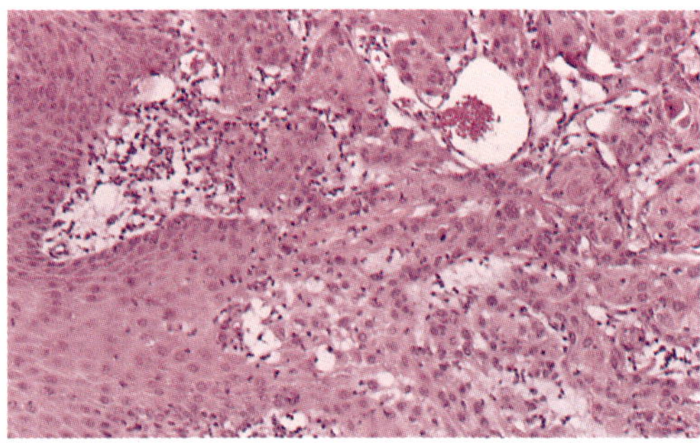

Fig. 13.4. *Histological appearance of a squamous cell carcinoma of the anal canal.*

uncornified (Fig. 13.4). About 90% of the perianal tumors and less than 50% of tumors of the anal canal are cornified. Ordinary squamous cell carcinomas make up about 80% of all tumors in the anal region.

Basaloid squamous cell carcinoma

Basaloid squamous cell carcinomas arise from cells in the transitional zone of the anal canal. The tumors are therefore also described as transitional or cloacogenic squamous cell carcinomas in American literature. Basaloid squamous cell carcinomas make up 20–30% of all malignant tumors of the anal region. Less than 10% of the perianal tumors have basaloid features.

Mucoepidermoid squamous cell carcinoma

These occur extremely rarely. They are seen in the anal canal and consist of squamous cells which produce both mucus and keratin. It is an anaplastic tumor form which has a poor prognosis with rapid development of distant metastases.

Division of squamous cell carcinomas into the above-mentioned histological types is often fortuitous and subjective; moreover, it depends on the site of biopsy. This, together with the fact that the classification is not thought to have great significance in the choice of treatment or the prognosis, causes the authors not to uphold the classification. Here, the tumors are merely defined as squamous cell carcinomas.

Epidemiology

Perianal squamous cell carcinomas are seen more frequently in men than in women (1.5:1), whereas the opposite is the case for squamous cell carcinoma in the anal canal. The tumors of both sites are seen at all ages, but the median age of presentation for both is about 62 years. There is no typical patient that is especially prone to the disease.

Etiology

The etiology is unknown. The tumors occur more frequently, however, in countries with poorer hygiene (e.g., India, Brazil) than in Western countries. In about 60% of the patients with perianal squamous cell carcinoma one or more accompanying benign diseases in the region occur, that is, condylomata, leukoplakia, anal fistulae or chronic fissures. Such coexisting diseases are, on the other hand, seen in only about 5% of the patients with squamous cell carcinoma of the anal canal.

Malignant transformation of condylomata acuminata to squamous cell carcinomata has been described.

Epidemiological and case-controlled studies seem to to show a relationship between venereal disease and the development of squamous cell carcinoma in the anal region, especially amongst homosexual men. A connection is thus thought to exist between syphilis and squamous cell carcinoma in the anal region, while the relationship between lymphogranuloma venereum and anal cancer is doubtful.

Squamous cell carcinoma of the anal canal has been described in patients with Paget's perianal disease.

Staging

The clinical staging of squamous cell carcinoma of the anal region varies from center to center. It has apparently been impossible to agree upon a common international system. The TMN-system has not been widely used, as it is difficult to distinguish tumors that are invading only the internal anal sphincter from those which are also involving the external anal sphincter, and also because spread to the rectum and/or the perianal skin region does not always worsen the prognosis. We, in accordance with the International Union Against Cancer, use the tumor's size and extent as basis for staging. Fig. 13.5 shows the size of the anal cancers on diagnosis.

Symptomatology

It is controversial whether the duration of symptoms (Fig. 13.6) is of significance for prognosis. In a large retrospective Danish study it could, however, be shown that the prognosis for the patient with squamous cell carcinoma of the anal canal was dependent on duration of symptoms (Fig. 13.7). Such a relationship was,

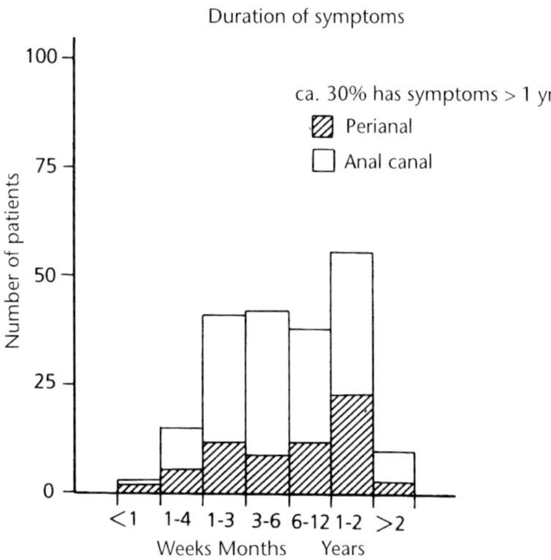

Fig. 13.6. *Histogram demonstrating duration of symptoms among patients with cancer of the anus before diagnosis.*

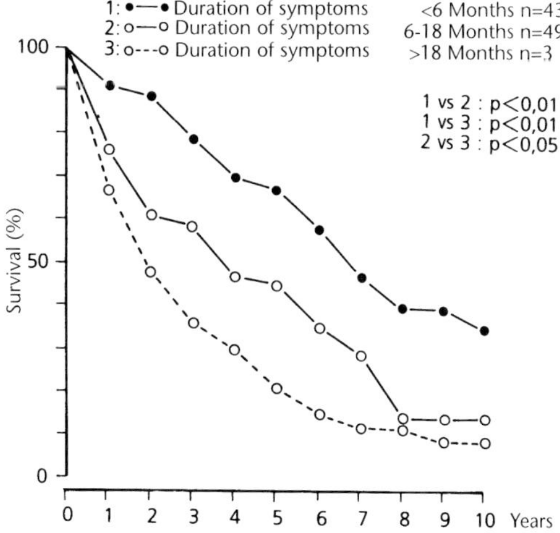

Fig. 13.7. *Survival curves (crude) for patients with squamous cell carcinomas of the anal canal in relation to duration of symptoms.*

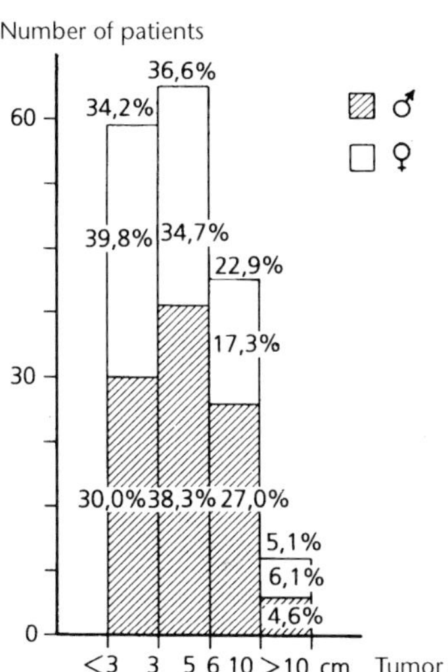

Fig. 13.5. *Histogram demonstrating the distribution of anal cancers in relation to tumor size and sex.*

on the other hand, not found amongst patients with perianal squamous cell carcinoma. It is important to emphasize that the average duration of symptoms before correct diagnosis was 9 months for patients with squamous cell carcinoma in the anal canal and 6 months for patients with perianal squamous cell carcinoma. These long periods of symptoms without diagnosis may be due to the patients' anxiety of a diagnosis which may lead to colostomy

Table I. Erroneous diagnoses in patients with anal cancer. About 40 percent are erroneously diagnosed.

Erroneous diagnosis	%
Hemorrhoids	38.4
Eczema	3.8
Anal fissure	10.6
Anal fistula	10.6
Anal abscess	9.6
Benign neoplasm	6.4
Negligence of anorectoscopy in patients with obvious anorectal symptoms	10.6

(patient's delay) and partly due to the doctors' lack of attention to these apparently benign symptoms (doctor's delay). Table 1 shows the most common misdiagnoses.

Significantly more patients with perianal squamous cell carcinoma present with a feeling of a mass in the region, while discharge from the anus (soiling), pruritus ani, and bleeding is more common amongst patients with tumors of the anal canal. Fig. 13.8 shows the symptoms in patients with anal cancer irrespective of site.

Diagnosis

Diagnosis is made from a precise history, a thorough anorectoscopy and most important by histological examination of pathological lesions of the anal region. It is important to emphasize that diagnosis is often made unexpectedly during routine examination of tissue removed from a benign disorder (hemorrhoids, fissures, fistulae). Fig. 13.9 shows a schematic representation of the diagnostic strategy for malignant anal disorders. CT-scanning should be used to assess the presence of lymph node metastases in the pelvis. The extent of the tumor in the perineum and anal canal can be assisted by ultrasound scanning (Fig. 13.10).

Differential diagnosis

The differential diagnoses are Bowen's perianal disease (intraepithelial carcinoma), Paget's perianal disease, condylomata acuminata, Buschke and Loewenstein's giant condyloma, histiocytoma, leukoplakia, malignant melanoma, basal cell carcinoma, Crohn's disease, chronic anal fistula, chronic anal fissure, keratoacanthoma, clear cell acanthoma, and lichen chronicus simplex et atrophicus.

Treatment
Local excision

Technique. The main indications for local excision of squamous cell carcinoma are perianal tumors and very small superficial tumors of the anal canal (less than 2 cm in diameter). Even

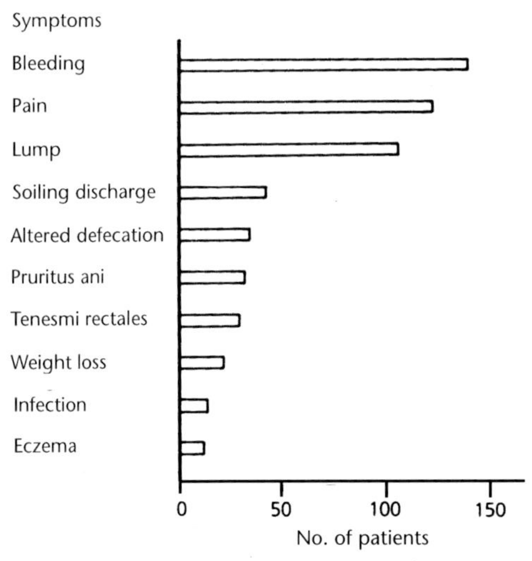

Fig. 13.8. *Most common symptoms in 176 patients with squamous cell carcinomas of the anus.*

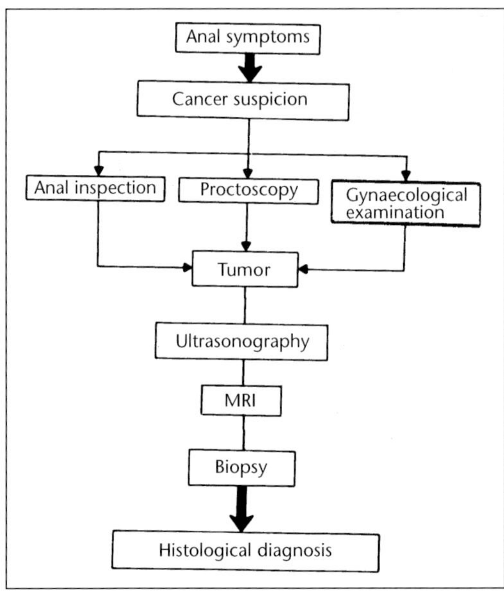

Fig. 13.9. *How to investigate a patient with suspected cancer of the anus.*

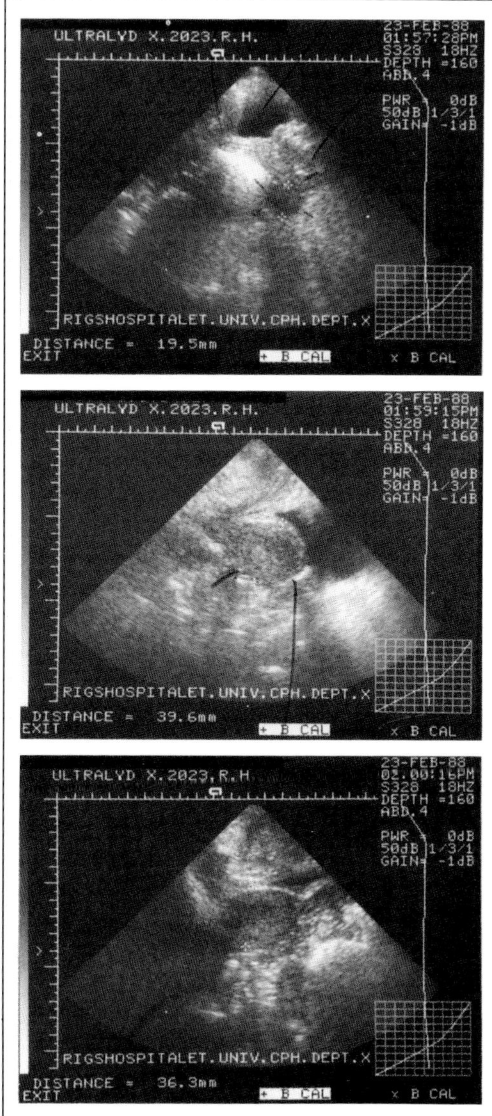

Fig. 13.10. *Ultrasonography of the perianal region demonstrating a squamous cell carcinoma of the anal canal.*

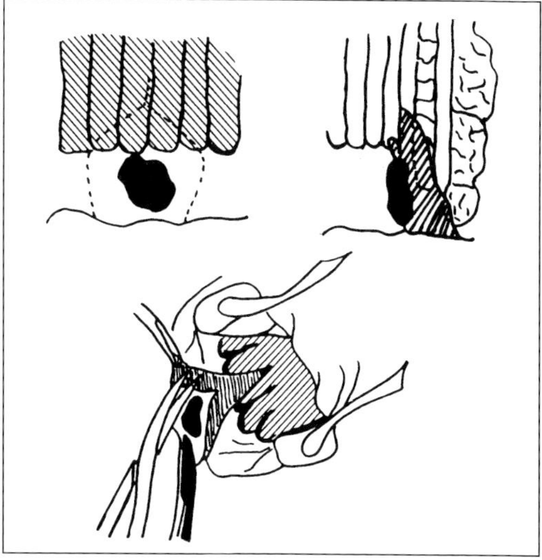

Fig. 13.11. *The principles of local excision of a small superficial anal cancer of the anal canal.*

hemostasis with diathermy. With very large wounds rotation flaps might be necessary to achieve closure.

Indication. Conventional local excision of squamous cell carcinomas should be employed only for perianal tumors that have not invaded the underlying muscle. Tumors in the actual anal canal should as a general rule not be treated by conventional local excision alone. If local excision of small superficial tumors of the anal canal is chosen the excision should be supplemented with subsequent rotation flaps or free skin grafts to achieve skin coverage. In most cases a temporary diverting colostomy will also be necessary. It is for these reasons that we prefer radiation therapy for these tumors.

Results. The results after local excision vary greatly and depend on the tumor's location and size. The recurrence rate is very high for tumors of the anal canal. The recurrence rate of perianal tumors treated by local excision is about 40% with a 5-year survival rate of about 65%. If underlying muscle is involved the incidence of lymph node metastases is about 60% and more radical treatment is required.

Abdominoperineal resection

Technique. The technique of abdominoperineal resection of the rectum is similar to the

with tumors located perianally it may be necessary to resect a small segment of the external anal sphincter and in a few cases also a part of the puborectalis muscle. In these cases Park's method using steel wire sutures can be used to reestablish continuity. Local excision is undertaken by making a circular or elliptical incision at a distance of at least 2 cm from the tumor (Fig. 13.11). All tissue is removed down to the internal anal sphincter which is thus exposed. The wound is left open after ensuring thorough technique used in the treatment of adenocarcinoma of the rectum except that the excision of the anus is more extensive than normal. This is because of the danger of local recurrence and because it may be necessary to use gluteal or perianal skin flaps to achieve wound coverage.

In women the operation is often extended by performing a posterior vaginectomy even if the rectovaginal septum is free of tumor. The latter point is, however, controversial.

Indication. The main indication for abdominoperineal resection of the rectum is tumors of the anal canal which do not disappear after radiation therapy alone or in combination with chemotherapy. All large (greater than 5 cm) squamous cell carcinomata of the anal canal should be treated in this manner whether or not there is total resolution of the tumor after radiation therapy. The large infiltrating perianal tumors should similarly be treated by rectal resection.

Results. The survival and recurrence rate after abdominoperineal resection of the rectum for squamous cell carcinoma in the anal region depends on the tumor's site, size, histological degrading, and local extend. The best results are achieved with small (less than 3 cm in diameter), highly differentiated tumors. The 5-year survival rate lies between 50 and 70% amongst patients that have been radically operated for squamous cell carcinoma of the anal canal. If, on the other hand, lymph node metastases are found at the primary procedure the 5-year survival drops to about 20% and if there are distant metastases survival is zero (Fig. 13.12).

Recurrence after abdominoperineal resection of the rectum occurs relatively late. In many cases these occur up to 8–11 years later.

Block dissection of lymph nodes

Technique. The technique for block dissection of inguinal lymph nodes is as follows:

1. A curved incision is made from the anterior superior iliac spine down over the pubic symphysis and up towards the opposite anterior superior iliac spine.

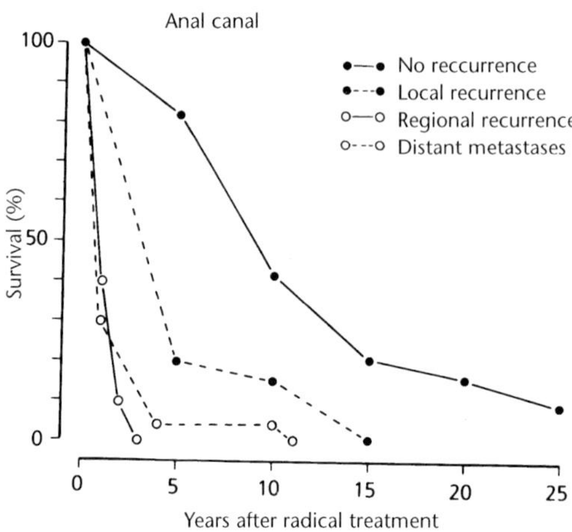

Fig. 13.12. *Crude survival curves of patients with squamous cell carcinomas of the anal canal in relation to the occurrence of metastases or not. (Treatment period 1943-1973).*

2. The skin of the area is cut away and all fat and lymphatic tissue is removed down to the aponeurosis and fascia down beyond the inguinal ligament on each side distally to the confluence of the long saphenous vein into the femoral vein. The superficial branches of the femoral artery are divided and ligated. All fatty tissue containing lymph nodes and lymphatic vessel is removed. The long saphenous vein is cut and ligated at its entry into the femoral vein.

3. The aponeurosis of the external oblique muscle is divided laterally from the external inguinal ring parallel to the inguinal ligament. Medially the incision is carried to the midline through the rectus sheath. The underlying muscle is divided so that the peritoneum is made free and retracted medially and proximally by an assistant.

4. The epigastric vessels are divided and ligated and the fatty and lymphatic tissue dissected free from the vessels up to the common iliac vessels. When the deep lymph node dissection has been completed the rectal sheath is closed and the transversalis fascia sutured to Cooper's ligament. The wound is now closed.

Indications. The significance of lymph node dissection in the pelvis is not clear, even though a few proctologists recommend it. The attitude to lymph node metastases in the inguinal region is, however, different. In such cases bilateral

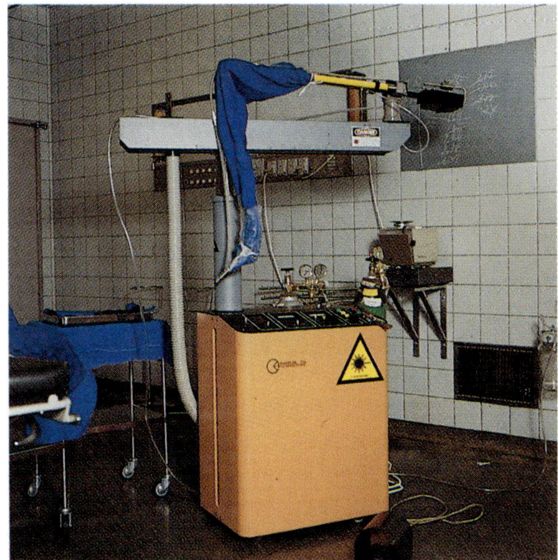
Fig. 13.13. *CO₂-laser apparatus.*

anal region has given surprisingly good functional results with long survival and low recurrence rate. Laser treatment is effective due to the facts that tumor cells are not spread because of the high temperatures that are reached in the interstitial tissue during treatment and that surgical dissection is improved because of diminished bleeding.

Technique. The CO$_2$ laser apparatus is used (Fig. 13.13). 30 watts are used for the skin and 20 watts for the subcutaneous tissue. The cutting beam has a diameter of 0.7 mm.

The tumor is excised at distance of 2 cm from the macroscopically visible tumor tissue. The mucosa is always included so that the external sphincter is revealed. Primary closure is achieved with the help of large or small plastic surgical reconstructions (Figs. 13.14 a–e). With the larger procedures a temporary diverting colostomy is employed.

Indication. Preliminary investigations show similarly promising results for the laser treatment of Bowen's and Paget's perianal diseases.

Results. Fifteen patients with a median observation time of 5.5 years have been treated with laser surgical excision. One patient developed recurrence after 8 months and a rectal resection was performed. Nine patients were operated on without colostomy, and six patients with colostomy. Only one patient has had his colostomy closed.

Radiation therapy

Radiation therapy has been used as treatment for squamous cell carcinoma for the last 70 years. During recent years the French oncologist Papillon has lead the way in this form of treatment. Earlier, the results were poor and accompanied by many complications, especially in the form of radiation necrosis and anal stenosis.

Newer methods of radiation have, however, limited the complications and more attention can now be paid to the sphincter-preserving method of treatment as an alternative to surgical treatment of tumors of the anal canal and the perianal region. Squamous cell carcinoma of the anal canal and its metastases are

lymph node dissection should be undertaken since 20% of these patients survive longer than 5 years. Prophylactic lymph node dissection of the inguinal region is, on the other hand, thought to be of no value.

Complications. Too vigorous undermining of the skin can cause necrosis. Bleeding.

Side-effects. Chronic lymph edema of the lower extremities.

Palliative surgical treatment (colostomy)

If the tumor is so extensive that excision or other forms of treatment are useless a diverting colostomy is the only treatment that can be offered.

Surgery for distant metastases

Distant metastases occur frequently with tumors of the anal canal. The commonest sites for distant metastases are the liver and lungs, but practically all organs may be involved. Solitary or well-delineated liver metastases can be treated by liver resection with good results.

Laser surgery

The use of CO$_2$ laser energy in the treatment of moderately large (2–6 cm), superficial non-infiltrating squamous cell carcinomas of the

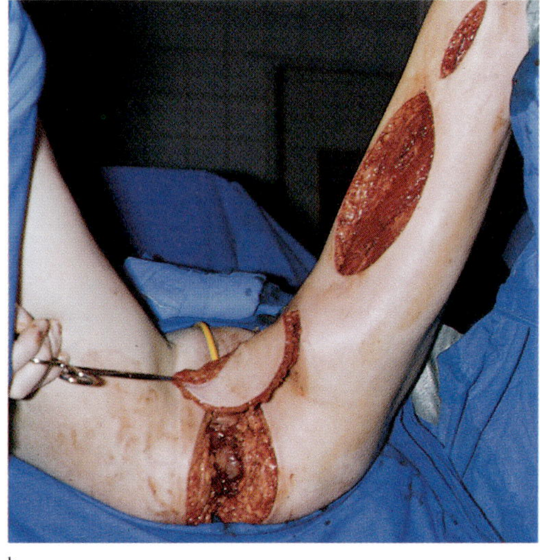

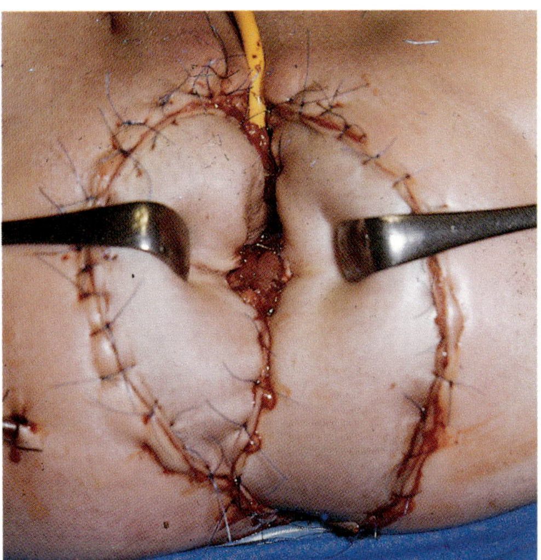

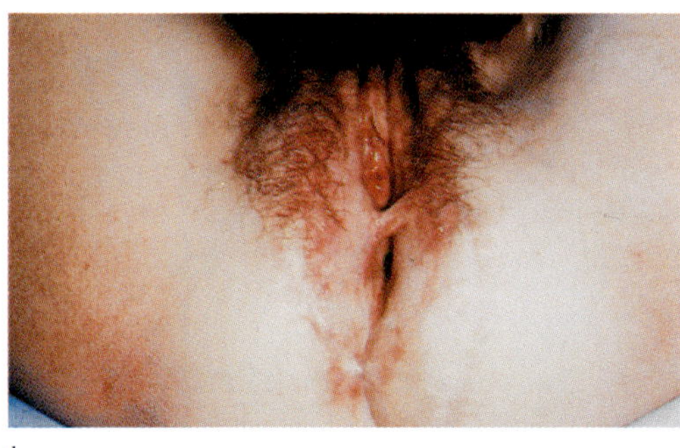

Fig. 13.14(a–c). *Superficial squamous cell carcinoma excised by laser surgery. Primary closure is achieved by musculo-cutaneous flaps. (d) Result after 7 years. The patient is continent.*

extremely sensitive to radiation therapy and even large tumors of 5–6 cm in diameter can be controlled by radiation therapy. Experienced clinicians from all over the world are of the opinion that radiation therapy is the best treatment for squamous cell carcinoma of the anal canal. Surgical treatment should perhaps be reserved for the treatment of any residual cancer still present 8 weeks after stopping radiation therapy. There are, however, no controlled studies concerning the different treating possibilities and it is still controversial which form of treatment is the best.

Radiation technique. Three different radiation methods are in use today, namely interstitial implantation of radium needles, external irradiation, or fractionated combined external and interstitial irradiation (split course irradiation). During the last two to three years we have used a combination of external irradiation with interstitial irradiation. With the introduction of pulse dose rate (PDR) equipment, where a single high activity source acts through catheters, a more individualised dose distribution and an elimination of radiation exposure to the staff can be achieved. The treatment usually consists of three-field external irradiation 46 Gy/23 fractions with five fractions a week to the anal canal and pelvic lymph nodes. One to four weeks later the tumor space is treated with 25.2 Gy PDR brachy-therapy with 42 pulses of 0.6 Gy, one pulse every hour. Seventeen patients have been treated and the results indicate that the treatment is highly effective but with substantial toxicity (necrosis).

Results. The results of radiation therapy should first be assessed after a few months since tumor regression may not be immediately apparent. This delayed response may be misjudged as radiation resistance and results in an unnecessary change in the treatment plan. The optimal effect is first seen 2 months after cessation of the radiation treatment.

Tumor of the anal canal

As with the other forms of treatment of squamous cell carcinoma it is difficult to evaluate the effect of radiation therapy since most series are small and heterogenously composed of patients with localized or advanced disease. The treatments are uncontrolled and often stretch over long periods and only a few authors have used actuarial analysis of their survival and recurrence data.

The 5-year survival varies from 32–79% and is accompanied by an incidence of complications of between 5–33%. Not all the complications are serious. Anal stenosis is now seen in only about 5–10% of cases. About 20% of the patients develop recurrence which later requires surgical treatment if this is technically possible.

According to Papillon, split course treatment is superior to external irradiation as far as curability and complications are concerned.

Perianal tumors

There are only few studies concerning the effect of radiation therapy on perianal squamous cell carcinoma since the standard treatment of these tumors is generally local excision. The results of radiation therapy as the only treatment for these tumors appears, however, to be just as good as the results of surgical treatment.

Complications of radiation therapy. Radiation necrosis with destruction of the perianal skin (Fig. 13.15), gastrointestinal problems (diarrhea, proctitis, bleeding).

Radiation therapy plus surgery

Surgical treatment with subsequent radiation therapy has been used for many patients with local or metastatic disease but it is doubtful whether this has resulted in longer survival or a better quality of life for the patient.

Cytostatic treatment

Chemotherapy has been used as the only treatment in a few cases. Active agents against squamous cell carcinoma are bleomycin, mitomycin C, and 5-fluorouracil. Chemotherapy is now used in many centers in combination with radiation therapy. It is, however, by no means certain that this combination therapy gives better results than, for example, radiation therapy alone, even though it has been postulated that cytostatic agents potentiate the effect of radiation therapy. The side-effects of

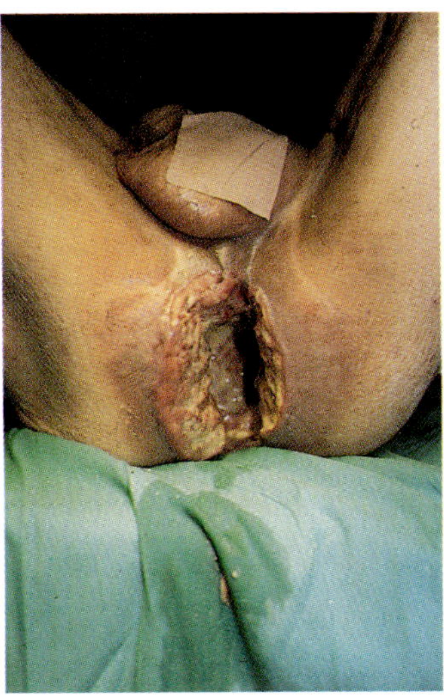

Fig. 13.15. *Necrosis of the anus and the perianal region after external irradiation of a squamous cell carcinoma of the anus. Biopsy from the wound edges demonstrated carcinoma despite treatment with 84,000 Gy.*

chemotherapy are considered to be a much larger problem so that centers in Denmark have been reluctant to employ combined treatments.

Combination treatment

Combined radiation and chemotherapy for the treatment of squamous cell carcinoma of the anal canal has been used especially in America. This has been used even if the tumors are infiltrating in an attempt to preserve the sphincter. It should be made clear that local or systemic recurrence is a high price to pay for the preservation of sphincter function.

Nigro (USA) has treated patients with external radiation (3000 rad [30 Gy], 300 rad per day) combined with the following chemotherapy: 5-FU 1000 mg/m^2 per 24 hours by continous infusion for 4 days commencing simultaneously with radiation therapy, mitomycin C 15 mg/m^2 on the first treatment day as a bolus and finally 4 days of 5-FU commencing 28 days after the initial treatment. This treatment was supplemented with surgical excision of the tumor if there were signs of residual tumor 2 months after commencing treatment, or on recurrence of the disease. Of 104 patients treated according to the above regimen there was an actuarial 3 to 5 years survival of 86% and 83%, respectively. Only 34 patients underwent radical surgical treatment in addition to the combined radiation and chemotherapy. Nigro (USA) thus recommend combined chemo- and radiation therapy to all patients with tumors in the anal canal and to patients with invasive perianal tumors. With large tumors (>6 cm) and with recurrent tumors treatment should be combined with abdominoperineal resection of the rectum. There were moderate side-effects in 99 of the 104 patients in the form of stomatitis, diarrhea, hair loss, and perianal dermatitis. The white blood cells were affected in fifteen patients, and five patients developed serious side-effects, although there were no deaths.

Prognosis

The prognosis depends among other things on the size of the tumor at the time of diagnosis, the duration of the symptoms, the histological differentiation, and the postoperative follow-up.

It is characteristic for the disease that it recurs frequently, even after an apparently radical treatment, and it is important that the patients are followed-up for a long period. In this connection it should be made clear that the recurrences can occur extremely late – up to 13 to 17 years after primary radical treatment (Figs. 13.16 and 13.17). The follow-up examinations to exclude local recurrences in the perineum as well as distant metastases, should thus include a full proctological examination test, X-ray, ultrasound scanning, and CT-scanning of the abdomen.

The actuarial 5- to 10-year survivals for patients with squamous cell carcinoma in the anal canal, irrespective of classification, dura-

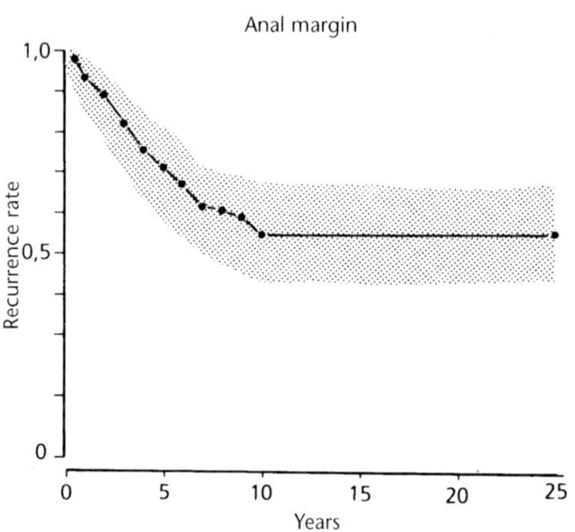

Fig. 13.16. *Recurrence rates after radical treatment of squamous cell carcinomas of the anal margin.*

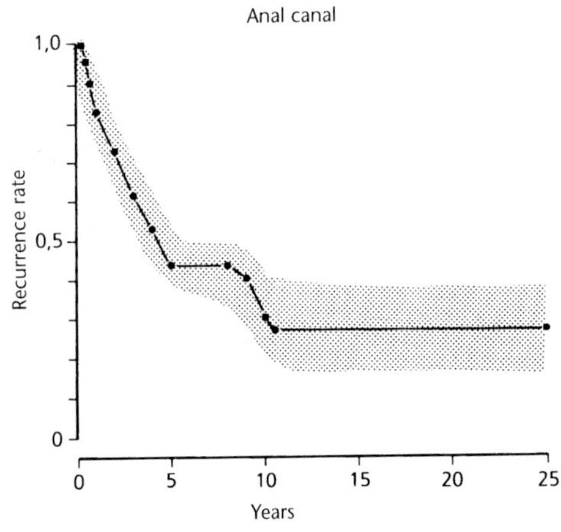

Fig. 13.17. *Recurrence rates after radical treatment of squamous cell carcinomas of the anal canal.*

tion of symptoms, and treatment, were 40 and 20%, respectively, in Denmark in the period from 1943 to 1973. The corresponding figures for perianal tumors are significantly higher at 66% and 45%. In the presence of regional lymph node metastases the 5-year survival is significantly diminished to about 20%. In the presence of distant metastases the survival is zero. These older Danish data which are in fact only of historical value are worse than present results and give food for thought. It should be remembered that patients were treated in a large number of different departments with no form of uniform treatment over a long period of time.

EXTRAMUCOSAL ADENOCARCINOMA OF THE ANAL CANAL

Definition
Extremely rare type of adenocarcinoma lying extramucosally, arising probably from the anal ducts. It should not be confused with the much more frequently occurring adenocarcinoma of the colorectal type arising from the columnar epithelium in the colorectal zone of the anal canal.

Pathological anatomy
The tumors are often large at the time of diagnosis, and it would be extremely difficult or even impossible to make the diagnosis on histological criteria.

They may present as an abscess-like process lateral to the anus in the ischiorectal fossa (Fig. 13.18) or as some other apparently benign perianal lesion (Fig. 13.19). Some divide the tumors into three types according to their site: anal, perianal (ischiorectal fossa), and fistula *in ano*. This classification is, however, controversial. Microscopically the diagnosis can only be made if the tumor is situated extramucosally. It presents as a typical adenocarcinoma (Fig. 13.20). Very occasionally one sees an anal duct with carcinoma *in situ* changes before transition to an invasively growing adenocarcinoma. The diagnosis is, however, clear from the histological point of view. The incidence of the disease is probably underesti-

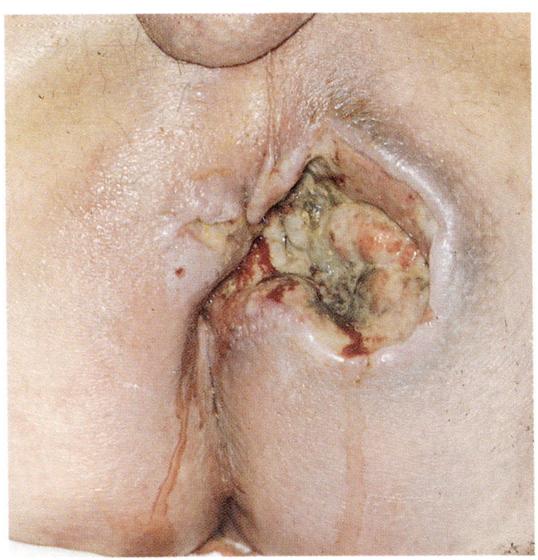

Fig. 13.18. *Huge necrotizing wound in the ischiorectal area: Microscopy demonstrated an extramucosal adenocarcinoma of the anus (anal duct adenocarcinoma).*

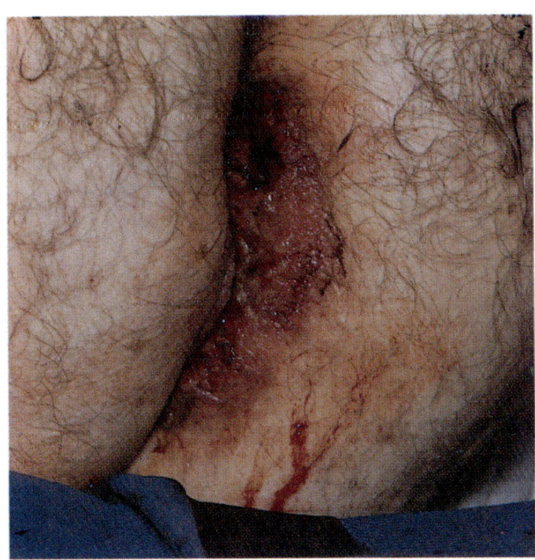

Fig. 13.19. *A patient with an indurated miscolored area to the left of the anus. Proctoscopy was normal. Biopsy demonstrated an anal duct adenocarcinoma.*

mated. Many cases may have invaded through the mucosa and are then classified as adenocarcinoma arising from the rectal mucosa.

Epidemiology
It is rare in young people. There is a slight preponderance amongst women. The incidence peaks in the seventh decade. Less than 100 cases have been reported in the literature.

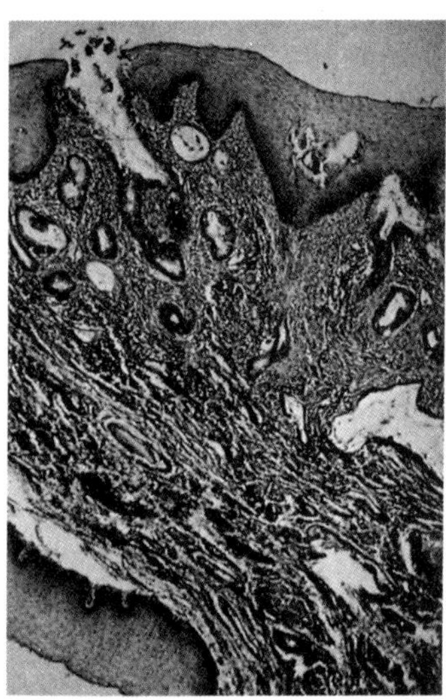

Fig. 13.20. *The histological appearance of an anal duct adenocarcinoma.*

The cases are often open to discussion since their documentation is first and foremost based upon clinical information while the histological evidence in many of the cases is weak. The etiology is unknown. The diagnosis is made in some cases on routine examination of excised hemorrhoids or anal fistulae.

Symptomatology

The most common symptom is a feeling of a mass followed by pain, bleeding, and tenderness as well as chronic secretion if the tumor is related to a long-standing anal fistula.

Diagnosis

It is important to be aware of this rare diagnosis in patients with anal abscesses which do not heal or in the presence of chronic anal fistulae. In the early stage, suspicion of the disease should be aroused if at rectal examination a tumor in the anal canal can be palpated without pathological changes in the mucosa since the process is initially extramucosal. All tissue removed, even that from clearly benign anal lesions, should be sent for histological examination. If there is a suspicious finding the diagnosis must be confirmed by biopsying the tumor. Anorectoscopy will not reveal extension of the tumor through

the mucosa. The tumor's extent can be assessed by CT or ultra-sound scanning of the pelvis and perineum. Ultrasound scanning and chest X-ray should also be carried out to exclude possible distant metastases.

Treatment

The treatment is surgical, and the only chance for cure is an abdominoperineal excision of the rectum with a wide perianal excision. Radiation therapy alone or in combination with chemotherapy is thought to have no effect on the course of the disease. Occasional small early tumors are cured by local excision, but the method cannot be recommended.

Prognosis

The prognosis is extremely bad. Most of the patients have distant metastases at the time of diagnosis and are dead within 6 months. In the Danish material the 1- to 5-year survival rate was 4.8%.

MALIGNANT MELANOMAS OF THE ANAL REGION

Definition

Highly malignant skin tumors arising from epidermal melanocytes in the skin of the anal region or anoderm. They may arise in junctional nevi or in completely normal melanocytes.

Epidemiology

This is an extremely rare site for these otherwise relatively common skin tumors. Just over a 100 cases have been described in the literature. They are extremely rare in young patients. They are seen most frequently in the fifth and sixth decades.

Pathological anatomy

Macroscopically the lesion often presents as a dark tumor of the anal canal (Fig. 13.21) or in the anal verge. The tumor may also, however, be amelanotic. Microscopically dermal inva-

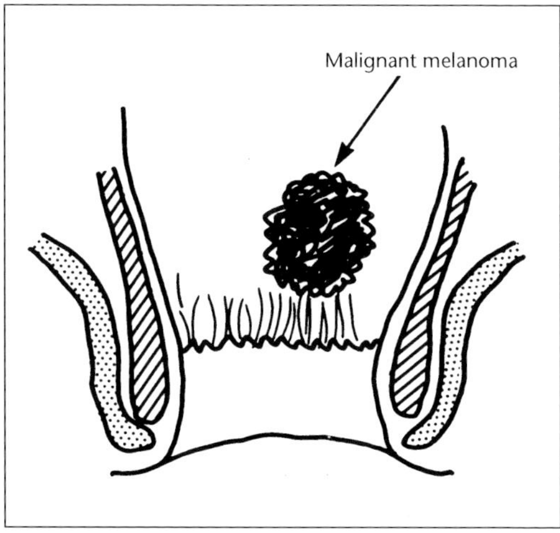

Fig. 13.21. *Sketch of malignant melanoma in the anal canal. The tumor can also be amelanotic.*

sion of atypical melanocytes from the overlying epidermis is seen. The atypical cells are often situated in nests at the epidermal–dermal junction. If atypical melanocytes in the epidermis are found alone without invasion of the dermis, this is described as an *in situ* malignant melanoma.

Symptomatology

Early symptoms: itching and a feeling of a mass. Late symptoms: ulceration and bleeding. In addition, poor general condition and weight loss because of distant metastases.

Diagnosis

The diagnosis is often made on routine microscopy of apparently normal removed tissue. When discolored lesions of the anorectum is encountered biopsy and microscopy should always be done. The diagnosis should not be made on frozen section. The biopsy should be representative and it may be necessary to carry out serial sections. It may occur that the diagnosis of metastatic malignant melanoma is made at some other site of the body without the primary tumor being apparent. In such cases the patient should undergo a thorough proctoscopic examination.

The patients should also undergo CT-scanning of the perineum and ultrasound scanning of the pelvis, perineum, and abdomen in a search for visceral metastases.

Differential diagnoses

Benign and malignant tumors of the anal canal as well as hemorrhoids.

Treatment

Abdominoperineal excision of the rectum with extended perineal resection. The tumor is not radiosensitive and chemotherapy is considered ineffective.

Prognosis

Poor. Most patients have distant metastases at the time of diagnosis. There are very few long-term survivors.

SARCOMA

Definition

Malignant connective tissue tumor arising from fat muscle or connective tissue in or around the anal canal.

Epidemiology

Extremely rare. Seen at all ages including children.

Pathology

The commonest form of sarcoma in the anal region is thought to be liposarcoma of the ischiorectal fossa followed by leiomyosarcoma of the internal anal sphincter and rhabdomyosarcoma of the levator ani muscle.

Clinical presentation

A mass in the region. The tumors are rather malignant. They frequently recur locally and there is a hematogenous spread.

Treatment

Excision of the rectum with wide perineal dissection.

Prognosis

Poor with few long-term survivors. The patients die of distant metastases.

ANAL MANIFESTATIONS OF MALIGNANT HEMATOLOGICAL DISORDERS

Definition

Anorectal lesion in connection with leukemia, malignant lymphoma including Hodgkin's disease.

Specific anorectal lesions occur rarely with chronic myeloid leukemia or acute leukemia. Symptomatic hemorrhoids, fistulae, and fissures may naturally occur as in the general population. On the other hand, leukemic infiltrations in the perianal skin are not infrequent in patients with lymphatic leukemia. Perianal abscesses and sepsis are frequently seen especially with the acute form of leukemia.

There is often severe pruritus ani with eczematous changes of the perianal skin in patients with Hodgkin's disease. Superficial or deep tumor infiltrations may also occur with this disease.

Treatment

When treating the anal lesions of patients with leukemia or malignant lymphoma one should be aware that these patients easily develop perianal abscesses after surgical procedures in this region. In particular incision and drainage of lesions resembling an abscess of the anal region in patients with acute leukemia can develop necrosis, sepsis, and bleeding. When a patient with the above-named blood diseases in the acute phase complains of hemorrhoids, fissures or fistulae, the treatment should be conservative. Only abscesses with clear fluctuation should be incised. For patients with lesser anal lesions we recommend warm sitz baths, laxatives, analgesics, and broad spectrum antibiotics. With the more chronic form the treatment of any anal lesions should also be as conservative as possible. The anal problems will often resolve spontaneously after the patient has been adequately treated with chemotherapy for their primary disease.

SUPPLEMENTARY READING

Bolivar JC, Harris JW, Brauch W, Sherman RT. Melanoma of the anal region. Surg Gynecol Obstet 1982; 154: 33.

Boman BM, Moertel CG, O'Connell MJ et al. Carcinoma of the anal canal: a clinical and pathologic study of 188 cases. Cancer 1984; 54: 114.

Greenall MJ, Quan SHQ, Urmacher C, DeCosse JJ. Treatment of epidermoid carcinoma of the anal canal. Surg Gynecol Obstet 1985; 61: 509.

Jensen SL, Nielsen OV. Cancer ani i Danmark 1943–1973. II. Sjældne former for perianale og anale tumorer Ugeskr Læger 1982; 144: 857.

Jensen SL, Hagen K, Shokouh-Amiri MH, Nielsen OV. Does an erroneous diagnosis of squamous-cell carcinoma of the anal canal and anal margin at first physicion visit influence prognosis? Dis Colon Rectum 1987; 30: 345.

Jensen SL, Shokouh-Amiri, Hagen K et al. Adenocarcinoma of the anal ducts. Dis Colon Rectum 1988; 31: 268.

Jensen SL, Hagen K, Harling H et al. Long-term prognosis after radical treatment for squamous-cell carcinoma of the anal canal and anal margin. Dis Colon Rectum 1988; 31: 278.

Klotz RG, Pamoukscoglu T, Souilliard DH. Transitional cloacogenic carcinoma of the anal canal: clinicopathologic study of three hundred seventy-three cases. Cancer 1967; 20: 1727.

Merlini M. Eckert P. Malignant tumors of the anus: a study of 106 cases. Am J Surg 1985; 150: 370.

Nielsen OV, Koch F. Carcinomas of the anorectal region of extramucosal origin with special reference to the anal ducts. Acta Chir Scand 1973; 139: 299.

Nielsen OV, Jensen SL, Petersen CF. Fejldiagnosticering af cancer ani. Ugeskr Læger 1976; 138: 723.

Nielsen OV, Jensen SL. Basal cell carcinoma of the anus – a clinical study of 34 cases. Br J Surg 1981; 68: 856.

Nielsen OV, Jensen SL. Cancer ani i Danmark 1943–1973. I. Carcinoma planocellulare – diagnose, behandling og prognose. Ugeskr Læger 1982; 144: 851.

Nigro ND. An evaluation of combined therapy for squamous cell cancer of the anal canal. Dis Colon Rectum 1984; 27:763.

Papillon J. Rectal and anal cancers. Berlin: Springer-Verlag, 1982.

Salmon RJ, Fenton J. Asselain B et al. Treatment of epidermoid anal cancer. Am J Surg 1984; 147: 43.

Welch JP, Malt RA. Appraisal of the treatment of carcinoma of the anus and anal canal. Surg Gynecol Obstet 1977; 145: 837.

XIV. Neuralgic conditions of the anorectum

These include the following conditions: coccygodynia, anorectal neuralgia, and proctalgia fugax. It is characteristic for these neuralgic conditions that no lesion can be found on physical examination to explain the pain. It is essential for the diagnosis to eliminate anorectal lesions as a cause for the symptoms. These conditions occur relatively infrequently, and the actual incidence is not known. Coccygodynia occurs most frequently (60%). The incidence of anorectal neuralgia and proctalgia fugax are 27% and 13%, respectively.

COCCYGODYNIA
Definition

Idiopathic anal or rectal pain of varying intensity, duration, and pattern that can be released by palpation of the coccyx or surrounding structures (muscle, fascia, ligaments).

Epidemiology

About 75% of the patients are women (40 to 50 years of age). The incidence is unknown.

Etiology

About 25% of cases can be related to trauma of the coccyx. It should be emphasized that fracture of the coccyx is not normally accompanied by coccygodynia. Contractures of the muscles covering the coccyx are given as the cause of coccygodynia. Many consider the syndrome as psychosomatic.

Symptomatology

The pain is localized either to the anus or the rectum. Some patients complain of regional spread of the pain. The pain is often moderate and varies from dull to moderate in intensity or to rhythmic pain contractions during defecation.

Diagnosis

The diagnosis is made from the history and negative anorectoscopy and a positive rectal examination with tenderness on palpation of the coccyx or its attached structures (muscles, ligaments, fascia).

Treatment

Various methods of treatment have been used: psychotherapy, massage of the region via the rectum, local anesthesia alone or in combination with steroid injections, coccygectomy (abandoned by all), radiation therapy (no effect).

The condition can best be helped by psychotherapy.

Prognosis

Unknown.

ANORECTAL NEURALGIA
Definition

An idiopathic painful condition of the anorectum that cannot be classified as coccygodynia or proctalgia fugax.

Epidemiology

As with the other neuralgias, about three-quarters of the cases occur in women of 50 to 60 years of age. The incidence is not known.

Etiology

It is often seen amongst women who previously have undergone gynecological operations (hysterectomy) and in men after surgery of the urine tract or anal canal.

Symptomatology

May be located in the anus or rectum. The intensity and duration of the pain vary, and the pain is migratory in about half the patients.

Diagnosis

Negative rectal examination and anorectoscopy. Previous anorectal operation or gynecological operation.

Treatment

Psychotherapy, morphines, sedatives.

Prognosis

Unknown.

PROCTALGIA FUGAX

Definition

Severe, intermittent, idiopathic, non-anal, rectal pain of short duration. The condition was first described in 1917, and Thaysen named it proctalgia fugax in 1935.

Epidemiology

The incidence cannot be stated. Is seen most often from 40 to 50 years of age, and about 75% of the patients are women. Single cases occurring in children have also been described. It is thought to occur commonly amongst doctors.

Etiology

Unknown, but many theories have been proposed: spasm of the levator ani muscles, sigmorectal intussusception, spasm of the hypogastric arteries (intermittent claudication), psychogenic causes.

Neither EMG changes in the striated muscles or intussusception of the sigmoid into the rectum have been demonstrated, however, since it is difficult to carry out the relevant investigations during an attack. Nothing abnormal has been demonstrated in these patients between attacks. The condition has probably a psychogenic background.

Symptomatology

The pain occurs characteristically in the morning, and patients are awakened by the pain. The pain occurs occasionally after defecation or prolonged sexual stimulation not ending in orgasm. The pain often radiates to the lower quadrant of the abdomen or the hip bones. The attacks of pain are severe and of durations varying from a few seconds to 15–30 minutes, typically 15–30 seconds. Occasionally the patient faints during an attack. The frequency of the attacks varies. Typically many months may pass between the attacks.

Diagnosis

Diagnosis is made from the patient's medical history, a negative rectal examination, and negative anorectoscopy.

Treatment

No specific treatment. Attempts have been made to treat the attacks with sublingual nitroglycerin, morphine, derivatives, sedatives, and psychotherapy. Often it is apparently enough to reassure a cancer-phobic patient that he or she is not suffering from a malignancy.

Prognosis

Good. In the authors' material, which consists of 25 cases followed-up for up to 7 years, twenty are asymptomatic and only five continued to have symptoms despite various treatments.

SUPPLEMENTARY READING

Boisson J, Debbasch L, Bensaude A. Les algies anorectales essentielles: etude clinique, pathogenique et therapeutique. Archives Francaises des Maladies Appareil Digestique (Paris), 1966.

Grant SR, Salvati EP, Rubin RJ. Levator ani syndrome; an analysis of 316 cases. Dis Colon Rectum 1975; 18, 161.

Neill ME, Swash M. Chronic perianal pain: an unsolved problem. J R Soc Med 1982; 75: 96.

Rockwood CA, Green DP. Fractures 937. Philadelphia: J.B. Lippencott, 1975.

Schuster MM. Constipation and anorectal disorders. Clin Gastroenterol 1977; 6: 643.

Thaysen EH. Proctalgia fugax, a little known form of pain in the rectum. Lancet 1935, ii: 243.

Thiele GH. Coccygodynia: cause and treatment. Dis Colon Rectum 1963; 6: 422.

XV. Pruritus ani

Definition

An unpleasant sensation of the perianal skin region that may be characterized by varying degrees of burning or itching, causing the patient to scratch the area. The condition may be divided into idiopathic and secondary types which may be acute or chronic. The two types cannot be distinguished from each other by the intensity of the symptoms, the macroscopic appearance of the perianal skin, sexual habits, bowel function, or profession. It is characteristic that there are marked daily fluctuations in the severity of the symptoms. Both conditions recur frequently.

Epidemiology

The condition is frequently seen by general practitioners, dermatologists, and surgeons. It occurs in all races, most frequently in the fourth to fifth decades. It is more common in women, who suffer from the secondary form, while men are most frequently affected by the idiopathic form. Patients eating a diet rich in fiber have a smaller chance of developing both idiopathic and secondary pruritus ani compared with a control group. There are more overweight patients in the group with idiopathic pruritus ani than in the group with the secondary form. In most studies over 50% of the patients suffer from idiopathic pruritus ani. In a prospective Danish study the cause of pruritus ani could be found in only about 65% of patients (secondary pruritus ani), while 45% of the patients were considered to have the idiopathic form of the disease.

Etiology

The cause of idiopathic pruritus ani is unknown, but many theories have been put forward. It is apparently induced by irritation of sensitive skin. Edema of the skin is not always apparent. The irritative condition often results in a vicious circle (Fig. 15.1). The patients scratch themselves causing injury to the superficial cells of the skin, whereby pathogenic microorganisms may cause an inflammatory reaction which accentuates the itch, and so on. Many factors predispose to an irritative condition in this region, i.e., large buttocks with accentuated dampness in the anal cleft and perianally, excessive sweating, and the use of inappropriate, nonporous underwear. Without doubt psychological factors also play a large role in many patients, i.e., stressful occupations or sexual neurosis. Such factors may be difficult to ascertain, so many of these cases are classified as idiopathic.

The secondary types may be divided into five main groups as shown in Fig. 15.2. The causes of secondary pruritus ani vary from practice to practice. Fig. 15.2 illustrates the situation in a specialized proctological practice. The picture will be completely different in a dermatological practice. In a specialist surgical practice the most common causes of secondary pruritus ani are anorectal disorders, such as prolapsed hemorrhoids, skin tags, fissures,

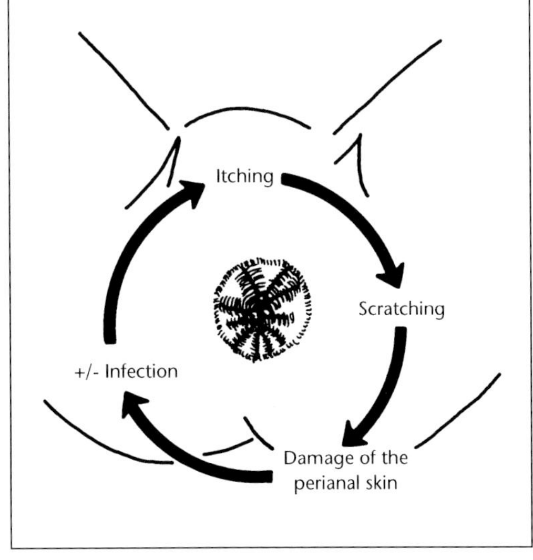

Fig. 15.1. *Vicious circle of idiopatic pruritus ani.*

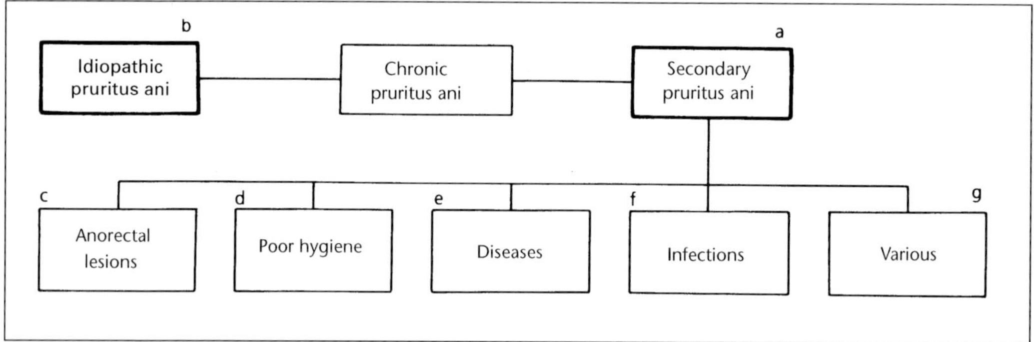

Fig. 15.2. *Classification of pruritus ani and causes of secondary pruritus ani seen in a proctological office.*

fistulae, etc. It should be emphasized that a pre-malignant or malignant condition may cause pruritus ani, although this is rare. Poor personal hygiene with smearing of fecal material in the region is often a cause. Sphincter disturbances after birth trauma or other forms of trauma (iatrogenic) may cause pruritus ani because of mucus discharge from the anus.

Dermatological conditions such as psoriasis, eczema, lichen simplex (neurodermatitis), neurotic excoriations or contact dermatitis often cause pruritus ani. It is important to remember contact dermatitis as a possible cause since many of the patients will have employed different skin preparations which can cause eczema, for example, antibiotic ointments, local anesthetics or steroid ointments or creams. Materials contained in the preparation of ointment, for example, lanolin, can cause contact dermatitis. Excessive use of water and soap in the anal region can cause irritation eczema.

One of the causes of pruritus ani in children may be prolonged exposure to wet, dirty diapers (nappy rash). Children are more easily infested with various worms, such as *Enterobius vermicularis*, which can cause severe itching.

Bacterial and/or fungal infections are seen in adults, for example, after steroid treatment or in association with diabetes mellitus or zinc deficiency and these infections may cause pruritus ani. Food allergy can cause frequent loose stools which can cause itching. It is unknown whether fruit allergy in the absence of accompanying diarrhea can cause pruritus ani.

Pathological anatomy

The macroscopic changes in pruritus ani vary from no changes to eczema and/or lichenization. Secondary infections with different fungi and/or bacteria are often found. Histologically hyperemia is found in the acute stage with mild parakeratosis and acanthosis in the more chronic stages (Fig. 15.3). Varying grades of infiltration with inflammatory cells are found.

Diagnosis

The diagnosis is made from a thorough history and physical examination that includes a dermatological examination, especially of the feet, joints, and face, supplemented with ano-rectoscopy, X-ray examination of the colon and a gynecological examination in women. Occupational details, information about allergy, especially fruit allergy, and an inquiry concerning the patient's underwear should be part of

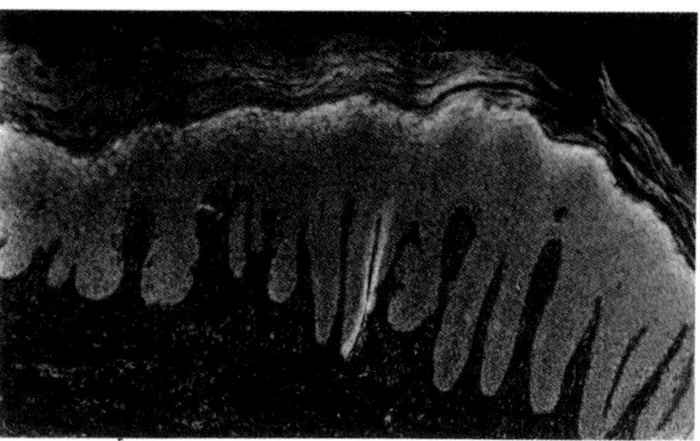

Fig. 15.3. *Histological appearance of idiopathic pruritus ani.*

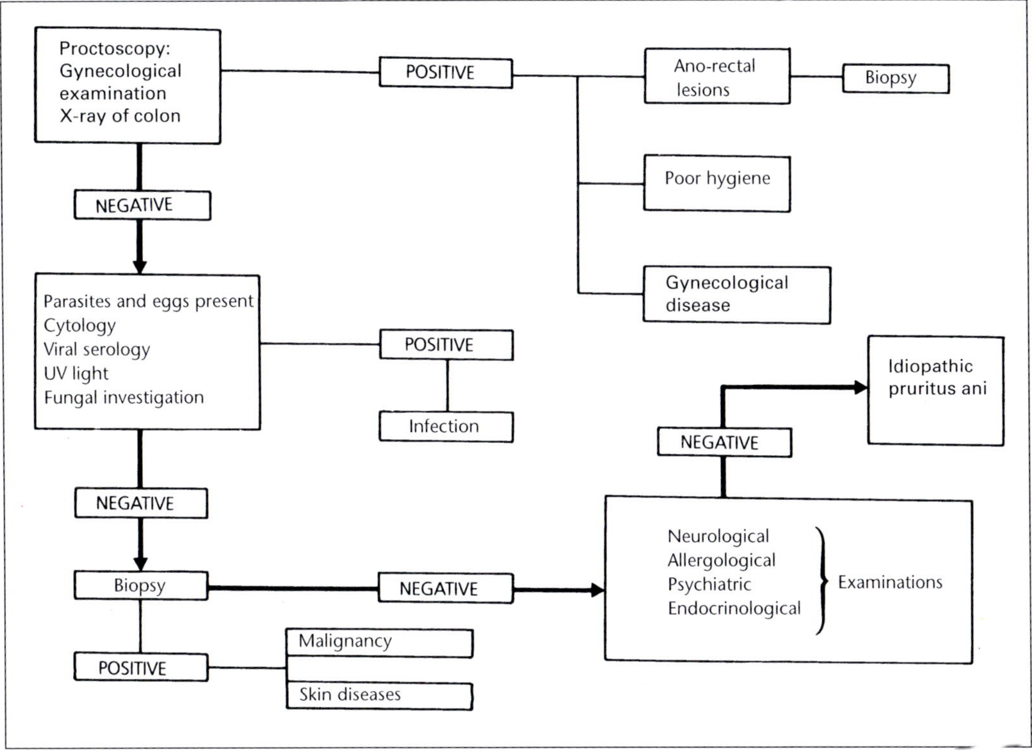

Fig. 15.4. *Flow diagram demonstrating the investigative procedures for patients with pruritus ani.*

the examination. A thorough microbiological investigation of the region should also be done (see Chapter 3). Wood's ultraviolet lamp is used for the diagnosis of *Corynebacterium minutissimum* (erythrasma). Fig. 15.4 shows a diagnostic flow-diagram used in the investigation of patients with pruritus ani. Biopsies of any visible lesions of the perianal skin should be taken for microscopical examination (Fig. 15.5).

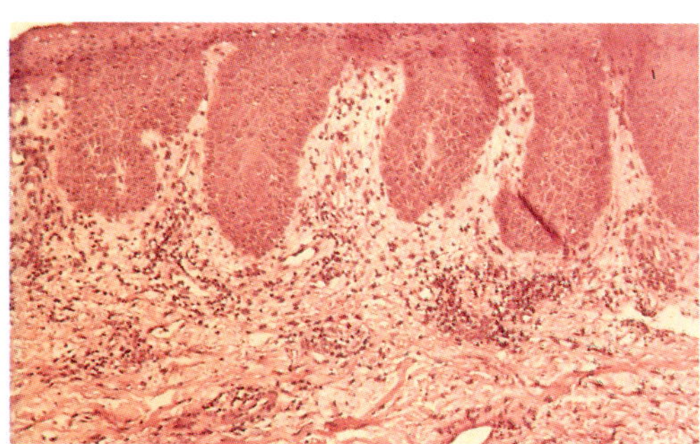

Fig. 15.5. *Histological appearance of zinc-deficiency induced pruritus ani.*

Treatment

Idiopathic pruritus ani

The treatment of pruritus ani depends on the basic cause. The treatment of the idiopathic pruritus ani is, therefore, particularly difficult. The treatment is only symptomatic, and it is often necessary to alternate from one treatment to another in an attempt to help the patient.

General guidelines

As a general rule, all medicine is discontinued with the hope that this may have an effect on the symptoms. In patients that have frequent, loose stools it should be ascertained whether this is caused by any specific food stuff. In the case of diarrhea this should be investigated according to general principles and antidiarrheal drugs, such as Retardin® or Imodium® may be tried. It is a general principle that overweight patients should reduce their weight and patients using inappropriate, non-porous underwear should be advised not to do

1. Reduce overweight
2. High fiber diet intake
3. Avoid foods initiating perianal itching
4. Avoid long visits to the toilet
5. Avoid coloured toilet paper
6. Avoid exaggerated cleaning of the perianal area, especially with soaps
7. Use a mild hygiene creme for cleaning of the perianal area after stools
8. Dry the perianal area with a hair dryer instead of toilet paper
9. Wash underwear with non-synthetic soaps
10. Avoid cortisones in the perianal area

Table 1. *Rules for patients with idiopathic pruritus ani.*

so. Patients are instructed concerning general principles of perianal hygiene. The authors' standard advice to patients with pruritus ani can be seen in Table 1. In addition to the above guidelines, patients are advised carefully to cleanse the anal region with ordinary water (possibly aided by a handshower if a bidet is not available) after each motion and every morning and evening. Alternatively, a cleaning agent such as Bidet® and a linen cloth may be used. Careful drying of the region is important; we advise patients to use an electric hairdryer on the warm setting for this purpose.

Local treatment

Possible local treatments include various cleansing agents, ointments or creams, injections, radiation therapy or surgery. We have experience only with dermatological preparations.

Cleaning agents, ointments, and creams. We have used Bidet® cleansing emulsion as a prophylactic agent in an attempt to prevent recurrent attacks. In an attempt to break the vicious circle of itching we have used either hydrocortisone or tumenolammonium containing creams. The incidence of recurrence is lower amongst patients treated with daily cleansing emulsion than the patients who received no treatment after a follow-up period of 12 months (Fig. 15.6). Bidet® cleansing emulsion is used morning, evening, and after each motion. Others have had success by cleansing the region with mucilago borica or cold cream.

We find that Tumedan cream gives better symptomatic relief than hydrocortisone cream when used for acute recurrences of pruritus ani with edema and irritation of the perianal skin.

The symptomatic effect of the two creams in non-acute recurrences of chronic pruritus ani is shown in Fig. 15.7. The authors' standard treatment of acute or chronic idiopathic pruritus ani is, therefore, Tumedan cream, since the side-effects of this preparation are less than those of steroid creams.

Long-term treatment of this area with powerful steroids should be avoided, since the risk of atrophy of the skin is great. Local treatment should be accompanied by general advice for patients with pruritus ani.

Other forms of treatment for idiopathic pruritus ani. For historical reasons the following should be mentioned, although the results are extremely doubtful:

1. Injection treatment: Either as sclerotherapy with hydrochloric acid or phenol in oil or injec-

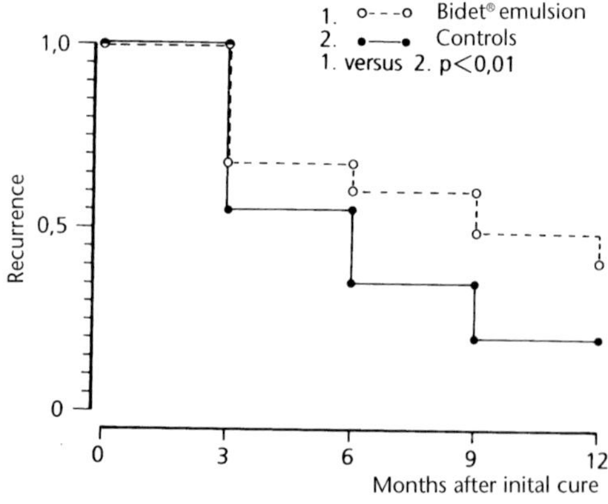

Fig. 15.6. *Recurrence rates of pruritus ani after treatment with Bidet*[R] *emulsion for 12 months compared to no treatment.*

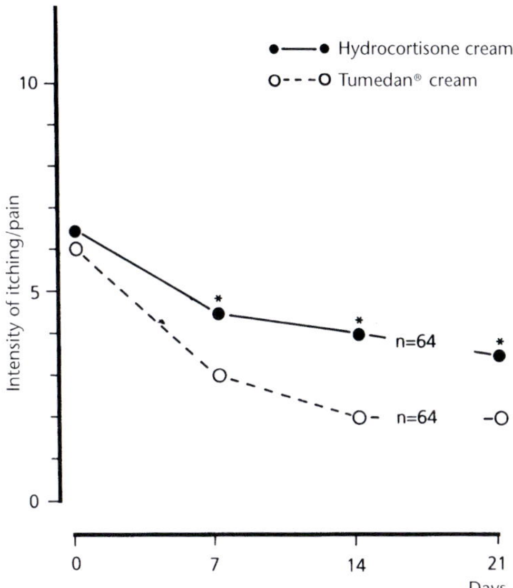

Fig. 15.7. *Effect of Tumedan® and cortisone cream on symptomatic relief in patients with pruritus ani.*

tion of long-acting anesthetics or alcohol causing cutaneous anesthesia.

2. Surgical treatment: The surgical methods that have been employed were Ball's operation (semi-ellipsoid incisions in the skin on each side of the anus with dissection of the external sphincter free from the skin and subcutaneous tissue and of the internal anal sphincter from the anoderm of the anal canal) or modifications hereof including excision of the perianal skin.

3. Radiation therapy: This has been attempted with varying effect. Serious burns of the skin have been described after radiation therapy.

Secondary pruritus ani

Anorectal lesions. Hemorrhoids, fissures, fistulae, abscesses, premalignant or malignant tumors (Figs. 15.8–15.10), sphincter lesions and so forth are treated as mentioned in the respective sections of the book. All these patients are, moreover, advised to follow the general principles pertaining to the management of anal irritation.

Poor Hygiene. Patients with poor personal hygiene are instructed to cleanse the anal region thoroughly after defecation. Cleansing emulsions can be used with advantage. Very hairy patients are advised to remove the hair from the region. Normally a local treatment such as Tumedan cream is used in the acute phase in order to break the vicious circle of itching.

Infectious pruritus ani

Fungal infections

The fungal infections causing pruritus ani are superficial mycoses of the skin and hair of the perianal region. They include dermatophytosis (*Trichophyton*, *Microsporum*, and *Epidermophyton*) and yeast mycoses (*Candida albicans*). Many preparations are available for local treatment. We often see patients with dermatophytosis being treated with nystatin, but it should be emphasized that this treatment is indicated for yeast mycosis. Substances such as sulfosalicylate ointment, sulfha paste, and methylroseanilinic spirits are often just as effective as commercial preparations in the treatment of dermatophytoses.

Many cases of intertrigo or eczema of the anal region may be caused by infection of yeast fungi since dampness, warmth, and damaged skin surfaces increase the risk of infection with

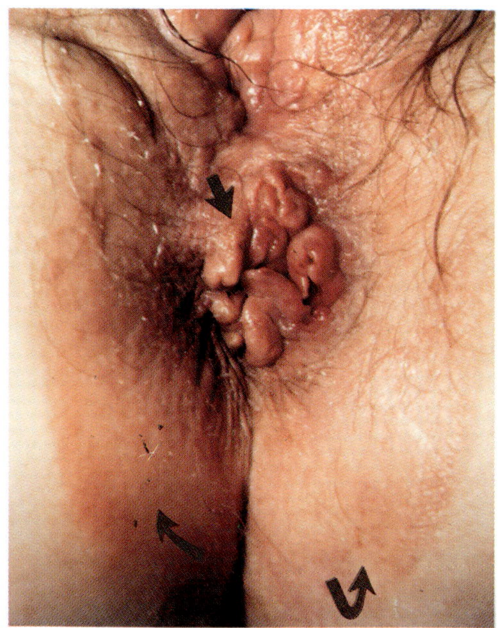

Fig. 15.8. *Patient submitted with severe pruritus ani and skin tags.*

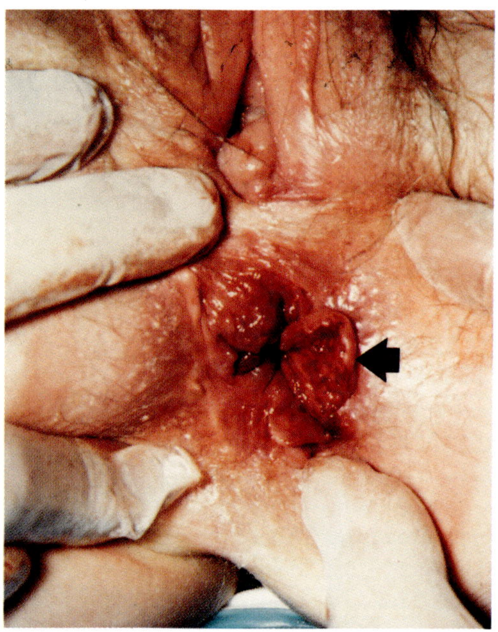

Fig. 15.9. *Same patient as in Fig. 15.8. Anal inspection demonstrates an indurated skin tag at 3 o'clock. Biopsy reveals squamous cell carcinoma. The patient has been treated with different ointments during the last 6 months before a careful examination was carried out.*

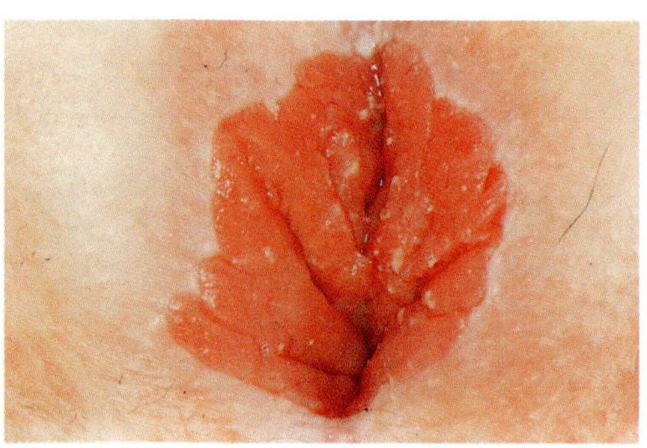

Fig. 15.10. *Patient treated for pruritus ani with different ointments for several years. Biopsy demonstrated an anal cancer (adenocarcinoma of the anal ducts).*

these fungi. They should be treated locally with nystatin (Mycostatin).

Bacterial infection

Impetigo is generally treated with antibiotic-containing ointment or cream applied a few times daily. Tetracyclin, Fucidin® or neomycin may be used.

Folliculitis may be treated with sulfa paste.

Erythrasma (*Corynebacterium minutissimum*) is treated by oral erythromycin. Local treatment in the form of sulfosalicylic acid ointment can also be used.

Virus infections

We generally treat condylomata acuminata by simple excision, but podophylline spirits may also be used. Herpes simplex of the anal region may be treated with Zovirax® tablets and/or cream.

Parasitic infection

Pediculosis pubis is overcome by treatment of the hairy region with Prioderm®. The treatment is repeated after 8 days. If the patient is very hairy the area should be shaved.

Scabies can be treated with 10% dixanthogen vaseline.

Skin diseases. Psoriasis is most often managed by some form of local treatment. Amongst other things steroid or tar preparations may be used.

Lichen simplex (neurodermatitis) is treated by psychotherapy supplemented with some form of local treatment either steroid or tar ointments. Oral antihistamines may be beneficial.

The same treatment is given to patients with neurotic lesions. Psychotherapy is very important here.

Contact dermatitis is treated by eliminating the toxic cause as well as local treatment in the form of short-acting steroid treatment or tar ointments.

Medical causes. If a medical condition is found to be the cause of chronic pruritus ani, this should be treated in the first instance. Until the primary disease is under control the general guidelines for the treatment of pruritus ani are employed.

Allergic patients can be treated with antihistamines.

In patients with diabetes mellitus (type 1) attempts should be made to establish normal insulin levels throughout the day. Patients with mature diabetes are given dietary supervision possibly supplemented with insulin or oral antidiabetic agents.

Patients with psychological diseases or problems are referred to psychiatrists for psychotherapy. Adipose patients are placed on weight-reducing diets.

Prognosis

The prognosis of patients with pruritus ani depends on the cause of the disease and its treatment. The common factor for both groups of patients is that the intensity of the itch varies greatly during a 24-hour period, and the recurrence rate is high after an apparent remission. Patients with idiopathic pruritus ani have the worst prognosis and their treatment is often a thankless task for the doctor.

Fig. 15.6 shows the recurrence rate for patients randomized to either prophylactic treatment with Bidet cleansing emulsion or standard cleansing of the region.

Fig. 15.11 shows the recurrence rate for the different causes of secondary pruritus ani after primary remission.

As is shown in the figures, recurrence is usual for patients with pruritus ani.

Recurrence may be caused by the previously mentioned factors, such as diarrhea, stress, exaggerated sweating or recurrence of anal diseases causing pruritus like hemorrhoids, fissures, and so on.

Prophylaxis

It is important to instruct the patient in general principles of hygiene and to explain that it is usual that the condition tends to recur, and that it should be treated promptly in order to break the previously discussed circle as soon as possible.

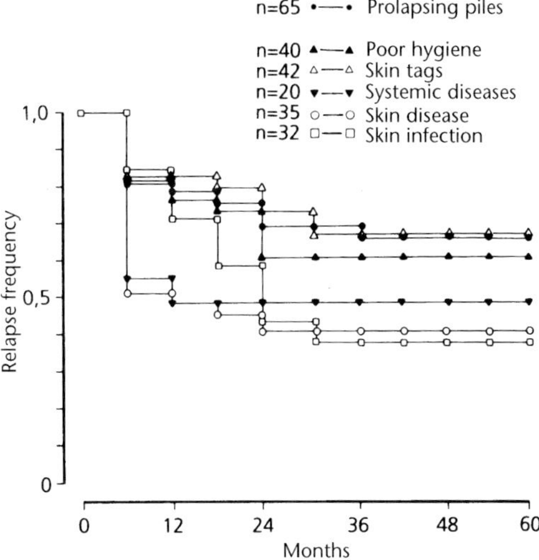

Fig. 15.11. *Recurrence rates of secondary pruritus ani in relation to cause after initial cure.*

SUPPLEMENTARY READING

Alexander S. Dermatological aspects of anorectal surgery. Clin Gastroenterol 1975; 4: 651.

Bahr Sir PH Manson: Manson's Tropical Diseases, 16th ed. London: Baillière, Tindall and Cassell, 1966.

Friend WG. The cause of treatment of idiopathic pruritus ani. Dis Colon Rectum 1977; 20: 346.

Goligher JC. Surgery of the anus, rectum and colon. 5th ed. London: Baillère Tindall, 1984.

Jellinek EH. Herpes zoster with dysfunction of the bladder and anus. Lancet 1976; ii: 1219.

Rook A, Wilkinson DS, Ebbing FJG, eds. Textbook of dermatology. Oxford and London: Blackwell Scientific Publications, 1969.

Sullivan ES, Garnjobst WM. Pruritus ani: a practical approach. Surg Clin North 1978; 58: 505.

XVI. Anal incontinence

Definition

A condition of reduced or absent ability to control the sphincter apparatus of the anal canal and/or muscles of the pelvic floor. One can distinguish amongst mild, moderate, and severe (total) incontinence. According to the cause, the condition can be divided into idiopathic or secondary anal incontinence. With mild to moderate incontinence the patient complains of varying degrees of soiling and occasionally incontinence for flatus or loose, watery stool. In severe or total incontinence control over the normal formation of feces is lost.

Pathogenesis

Most patients with anal incontinence display no abnormality in the anorectum on clinical examination, nor do they have any sign of neurological or inflammatory disease; these are grouped as idiopathic cases. Studies have shown that many of these patients have abnormal manometric or EMG patterns. Histological examination suggests that the cause is a partial denervation of the muscles. The histological changes are most marked in the puborectalis muscle and the external anal sphincter. The patients are often suffering from the descending perineal syndrome which can result in damage to the distal nerve branches of the pelvic floor muscles. These nerve branches become overstretched by recurrent and prolonged straining during defecation resulting in varying degrees of denervation damage of the muscles.

The puborectalis is the most important muscle in the maintenance of normal anal continence. The muscle is innervated by the motor branches of S3 and S4. Complete anal incontinence is nearly always due to weakness of the puborectalis, whereby maintenance of the normal anorectal angle is impossible. Damage and weakness of the puborectalis muscle is most often the result of trauma. Traumatic damage of the internal anal sphincter by anal dilatation, sphincterotomy or complete rectal prolapse, amongst other things, usually causes mild to moderate incontinence.

Older patients with mild to moderate degrees of anal incontinence are often suffering from fecal impaction (Fig. 16.1). This may cause an increased activation of the rectosphincteric reflex which normally causes the internal anal sphincter to relax when the rectum is distended. Overaction of the reflex can cause defective closure of the anal canal. This may result in leakage of thin stool, especially when intraabdominal pressure is raised as, for example, when coughing. Fecal inpaction is often seen in patients who have a reduced awareness of the degree of rectal filling (reduced rectal sensation), for example, patients with senile dementia, cerebrovascular diseases or sclerosis. Diagnosis is often missed because of routine emptying of the bowel before rectosocopy.

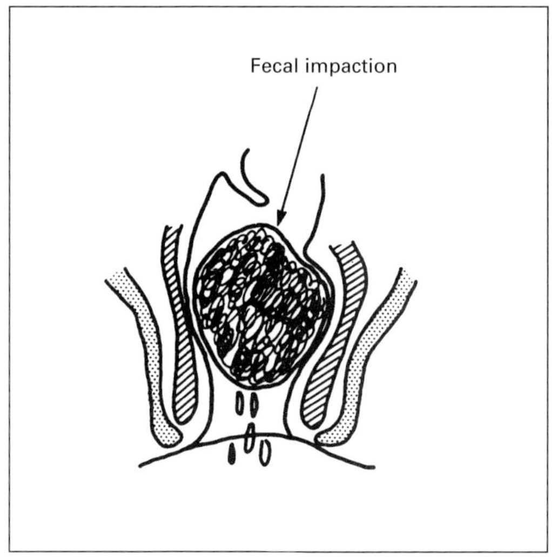

Fig. 16.1. *Fecal impaction causing moderate anal incontinence.*

Etiology

There are many causes of anal incontinence (Table 1). The incontinence is often due to trauma, anal diseases, neuromuscular disorders, inflammatory conditions of the bowel or weakness of the sphincter muscles because of old age. If the cause of incontinence cannot be found it is regarded as idiopathic.

Trauma

The trauma may be obstetric, operative (iatrogenic) or accidental.

Obstetric trauma. Perineal tears occur commonly during labor. In the mildest forms the tear extends towards or up to the anal sphincter whereby the superficial layers may be damaged. In the severe form complete rupture of the anal sphincter as well as tearing of the rectum mucosa occurs. Obstetric trauma of the perineum can also be accompanied by damage of the nerve supply of the sphincter muscles and pelvic floor musculature and cause anal incontinence without damaging the muscle itself.

Operative trauma. Transsection of the anorectal ring (puborectalis muscle) during operations for fistulae generally causes severe anal incontinence. Treatment of fissures with anal dilation (beware of eight-finger dilatation) or subcutaneous internal lateral sphincterotomy can cause incontinence. This form of incontinence is more common after anal dilatation than after sphincterotomy, even if the dilatation has only been 3–4 fingers. It is controversial whether hemorrhoidectomy *per se* can cause incontinence. On the other hand, if a gutter of rectal mucosa is pulled down to the anoderm during healing after a hemorrhoidectomy, leakage from the anal canal can occur.

Sphincter-saving resection of the rectum. Low anterior resection of the rectum or transsacral rectal resection (Mason) usually causes transient anal incontinence. The construction of an ileo- anal pouch or a coloanal anastomosis can be accompanied by temporary or permanent incontinence, especially at night. For many years normal rectal sensation and reflex activity of the anal sphincters after rectal distension have been regarded as essential for normal continence. Preservation of at least 6–7 cm of anorectum was considered necessary for the maintenance of continence. There is, however, clinical evidence that the sensory nerves which are responsible for normal rectal sensation and the reflex inhibition of the anal sphincters in response to rectal distension lie extrarectally. This is supported by the fact that patients who have had the greater part of their rectum removed as a result of carcinoma still maintain most of the normal physiological mechanisms of continence.

Accidental anal trauma. Impaling accidents may sever or tear the sphincter apparatus totally or partially. Multi-traumatized traffic accident victims, especially those with severe pelvic fractures, sometimes have rupture of the perineum with tears of the sphincter apparatus.

Hemorrhoids

Patients with long-term prolapse of third or fourth degree hemorrhoids are often incontinent for flatus and loose stool. This is probably due to resistant mild dilatation of the internal anal sphincter musculature. The incontinence is often cured by hemorrhoidectomy.

Table I. Causes of anal incontinence

Normal function of the sphincter muscles
Diarrhoea
    Infections
    Inflammatory bowel diseases
    Metabolic disturbances
    Small bowel resections

Fistulae
Colostomy

Abnormal function of the sphincter muscles

Mild to moderate incontinence:
    Idiopathic
    Iatrogenic
    Rectal prolapse
    Age
Severe incontinence:
    Congenital abnormalties
    Trauma
    Complete rectal prolapse
    Anorectal malignancies
    Infections
    Neurological diseases

Rectal prolapse

Complete rectal prolapse is accompanied by incontinence because the prolapsed rectal mucosa interferes with closure of the anus supplemented by dilatation of the internal anal sphincter.

Inflammatory bowel disease

The inflammatory changes in the rectum in ulcerative colitis or less often Crohn's disease can reduce rectal compliance, and thereby cause anal incontinence. Anal incontinence in the presence of inflammatory bowel disease may also be due to destruction of the sphincter muscles as a result of inflammation. Many of the sexually transmitted anal diseases (see Chapter 9) can cause anal incontinence by the same mechanism (lymphogranuloma venereum, syphilis, gonorrhea).

Anal cancer

Any of the malignant anal tumors can cause anal incontinence by destruction of the external anal sphincter by tumor infiltration (Chapter 13).

Neuromuscular diseases

Neuromuscular disorders, massive infections of the anal region, vascular diseases and demyelinizing conditions of the spinal cord are often accompanied by anal incontinence. This form of anal incontinence is more serious than the others mentioned above since it is more difficult to treat these diseases adequately.

Anal incontinence of old age

Atrophy and weakness of the sphincter muscles resulting in incontinence is seen with increasing age. Not all old people develop incontinence, and it is unknown what factors predispose to this form of incontinence.

Anal incontinence in the presence of normal function of the sphincters and pelvic floor musculature

As shown in Table 1, severe infectious diarrhea can cause transient anal incontinence despite a normally functioning sphincter apparatus. Other similar conditions are also presented in Table 1.

Symptomatology

Mild anal incontinence

Control of formed and liquid feces, but occasional incontinence for flatus.

Moderate anal incontinence

Periodic incontinence of both flatus and formed stool.

Severe incontinence

Constant incontinence of both flatus and formed stool. The patient is forced to wear a diaper.

Soiling because of poor anal hygiene should not be regarded as anal incontinence.

Diagnosis

The degree of incontinence is judged from a thorough medical history. Does the patient use a diaper? Is the patient incontinent during sleep? What are the provoking factors (coughing, sneezing, lifting weights)? Investigations should include anorectoscopy and rectal examinations, as well as X-ray examination of the colon to exclude tumors, inflammatory conditions or spurious diarrhea.

The rectal examination is important in patients with incontinence and should include the sphincter's resting tone, voluntary contraction, and anal reflex. The presence of scar tissue, fistulae, hemorrhoids and skin tags should be noted. Manometry may also reflect something of the function of the internal anal sphincter. The value of manometry is, however, debatable. Resting-pressure and maximal squeeze-pressure are lower in these patients than in controls. It should be known that great variation in pressure measurements occur in the individual patient. This may lower the evaluation of the effect of operative treatment for anal incontinence. EMG is used to assess the external anal sphincter. The investigation is considered more important than anal manometry. The measurement of the anal angle is undertaken by a dynamic X-ray examination, a so-called defecography. This investigation can also reveal descent of the pelvic floor during straining.

Treatment

The presence of moderate or severe anal incontinence is an indication for surgical treatment whereas mild incontinence can be treated conservatively by squeeze excercises, changes in diet, and/or improved hygiene. Cases of incontinence in the presence of normal sphincter function are also treated conservatively.

Many different operative methods are available. The choice of method depends on the patient's age, the experience of the surgeon, the type and cause of the incontinence. The operative methods that will be discussed, also include three perianal procedures for the treatment of complete rectal prolapse: anal sphincteroplasty, postanal repair, gracilis transposition, Thiersch's ring, Delorme's operation, rectosigmoidectomy, and artificial sphincter.

Anal sphincteroplasty

Indication. Traumatic sphincter lesions where the muscle mass has been preserved and the sensory and motor innervation is intact. Incontinence secondary to diseases of the central or peripheral nervous system cannot be treated by this method. The method requires the presence of scar tissue. It is best instituted 3–6 months after the trauma, when adequate scar tissue has been formed.

Preoperative investigations and treatment. Digital assessment of the muscle mass and ability of voluntary muscle contraction, possibly supplemented with EMG.

Preoperative bowel cleansing is performed and perioperative antibiotics given. The patient is placed in the lithotomy position with the Trendelenburg tilt.

Technique

1. An anterior circumanal incision is made 1 cm external to the mucocutaneous junction.
2. Mucosa is dissected free from the internal and external anal sphincter creating a tube of mucosa.
3. External anal sphincter together with fibrotic scar tissue is dissected free in the area between the muscle and perianal fatty tissue.
4. The scar tissue is transsected at its midpoint.

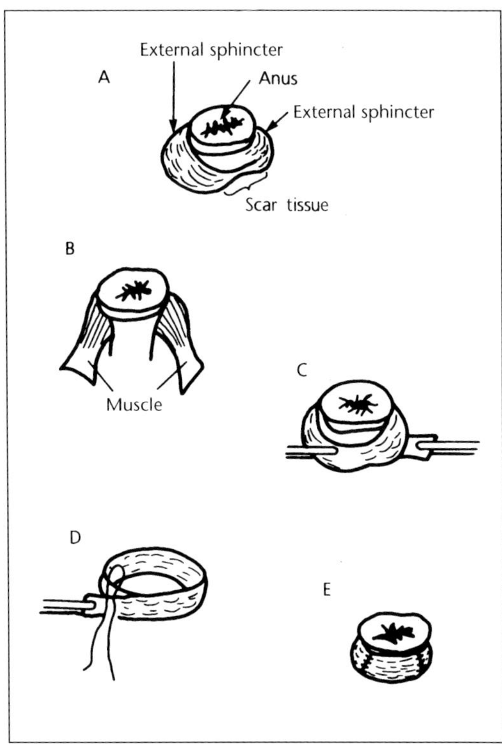

Fig. 16.2   *The principles of sphincteroplasty.*

5. The two ends of the muscle and scar tissue are positioned so they overlap each other without tension around a finger in the rectum.
6. The ends are sutured to each other in this position with 2-0 absorbable suture materials.
7. The incision in the anoderm is loosely closed with prolene sutures.

Hemostasis with diathermy is carried out taking care to avoid damage to the branches of the pudental nerve that run posteriolaterally into the sphincter muscles. The steps of operation are shown in Fig. 16.2.

Postoperative management. The patient is put on a liquid diet for the first 2–3 days after the operation, followed by dietary fiber supplement and a full diet from the fourth postoperative day. The patient is examined daily for bleeding and infection. The patient is discharged after normally controlled defecation 7–10 days after surgery and checked 3–4 weeks later by rectal examination and possibly EMG.

Prognosis. The results depend on the duration of the sphincter lesion, age of the patient, and whether there have been previous attempts at sphincter reconstruction. The older the patient, the longer the duration of lesion, the greater the number of previous attempts at reconstruction, and the worse the result.

About 50% of patients achieve normal continence, and 30–40% experience a significant improvement with only occasional incontinence for flatus. About 20% of the patients show no improvement after the operation.

Postanal repair

Indication and background. This is used in two groups of patients with poor function of the anal sphincters, namely idiopathic incontinence of old age and rectal prolapse. The patients are often women with minimal external anal sphincter function and weak voluntary contraction. The puborectalis muscle becomes elongated and slack, especially posteriorly, and the anorectal angle is more or less obliterated. The aim of postanal repair is to reconstitute the anorectal angle, increase the length of the anal canal and the flap-valve-apparatus.

Preoperative treatment. As for anal sphincteroplasty.

Technique (Fig. 16.3)
1. Semicircular incision behind the anus in the intersphincteric groove (A).

2. The external anal sphincter is cut free from the internal anal sphincter by sharp and blunt dissection, until the puborectalis muscle is reached (B).

3. Waldeyer's fascia, which binds the rectum at this point, is incised, and the supralevator space is entered. Levator ani muscles can be identified on each side (C and D).

4. The levator ani muscles are approximated on each side with 2-0 absorbable sutures. The

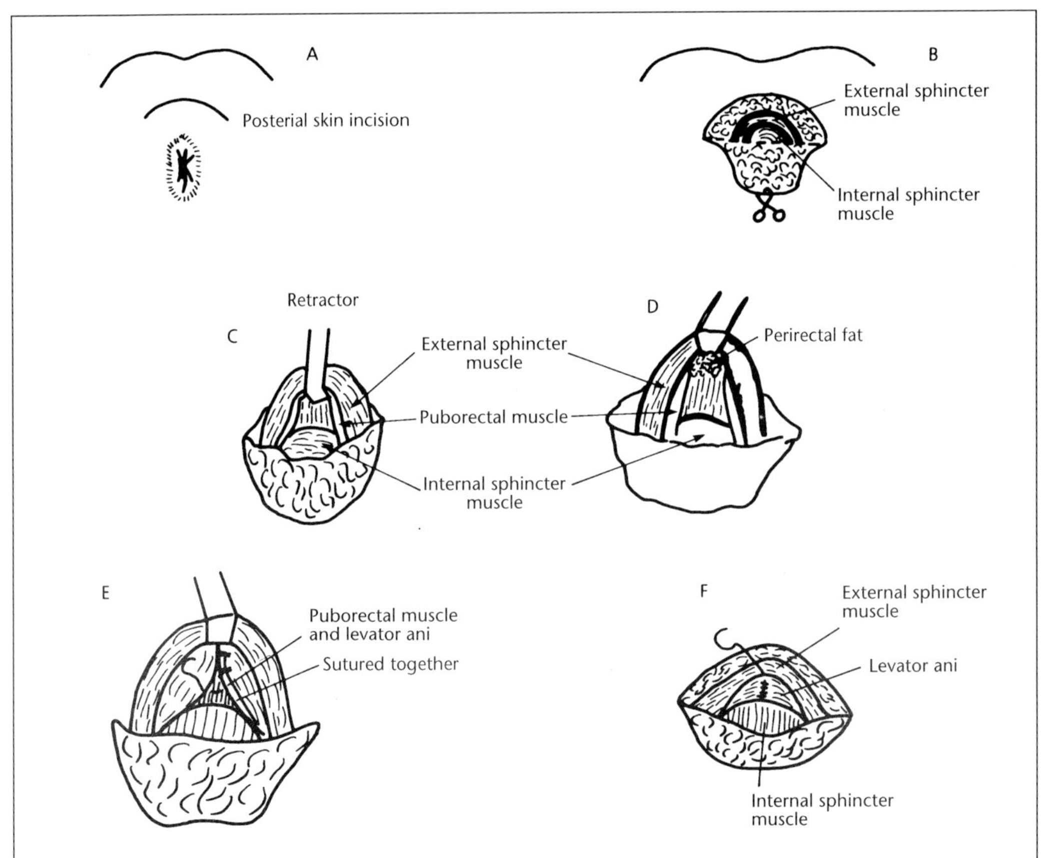

Fig. 16.3. *The principles of postanal repair.*

same is done to the puborectalis muscle and the external anal sphincter (E and F).

5. The skin incision is loosely closed with prolene sutures after insertion of a fine suction drain (size 14).

Postoperative treatment. As for anal sphincteroplasty. Patients undergoing postanal repair should, however, avoid defecation and be kept constipated for about 10 days. To avoid straining on defecation glycerin suppositories can be given daily so that defecation occurs by contraction of the rectal wall.

Complications. Wound infections, necroses in the sphincter due to sutures which have been too tightly placed, rectal inertia.

Results. The results are reported as good. More than 70% of the patients will become continent for both solid and liquid feces, and about a further 10% for solid feces only. This is after an observation period of up to 6 years.

Gracilis transposition

Indication. Can be used after extensive loss of sphincter musculature following trauma or infections, after many other attempts at reconstruction of the sphincter apparatus have failed. Not suited to elderly people or patients with neuromuscular diseases.

Preoperative treatment. The patients undergo complete bowel preparation. The patient is placed in the lithotomy position (Fig. 16.4) to provide easy access to the thighs and anus. Preoperative antibiotic cover is instituted.

Technique (Fig. 16.4)

1. Three incisions are made on the medio-aspect of the thigh proximally, half-way down, and at the level of the knee. The gracilis muscle

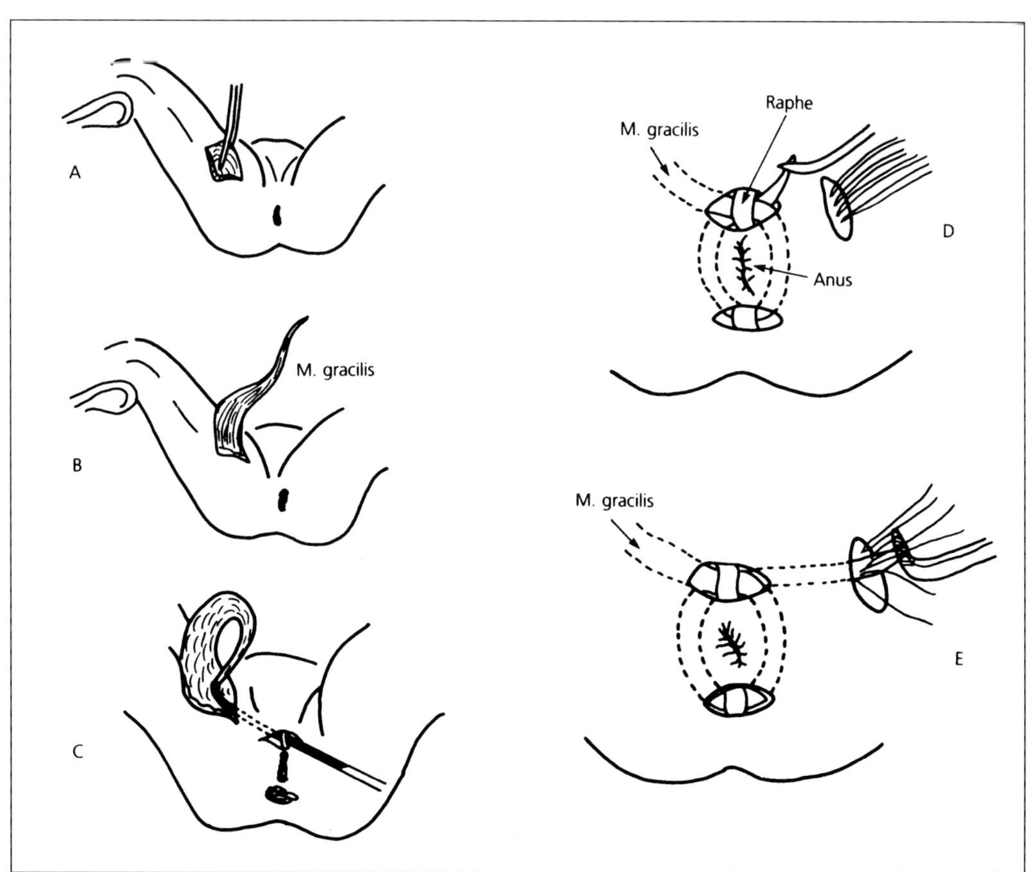

Fig. 16.4. *The principles of gracilis muscle transposition.*

is identified in the proximate incision, and a rubber band is placed around the muscle. The dissection is continued proximally to the vascular and nerve pedicel.

2. The muscle is dissected free via the two other incisions and by blunt dissection between the incisions. The muscle is freed as far distally as possible through the distal incision, and it is transsected at the level of its ligament. The two distal incisions are closed.

3. Incisions are made about 1.5 to 2 cm anterially and posteriorly to the anal verge. By blunt dissection a tunnel is made between proximal thigh incision and anterior anal incision and the gracilis muscle is pulled through.

4. An additional tunnel is created to lie in the extrasphincteric space on each side of the anal canal, and the gracilis muscle is placed 360 degrees around the anus in a clockwise direction, if the muscle from the right side has been used, otherwise the muscle is placed in an anticlockwise direction. The proximal thigh incision is closed.

5. An incision is made over the contralateral ischial tuberosity. Three non-absorbable sutures are placed in the gluteal fascia. The tendon is pulled through a tunnel between this contralateral incision and the first anal incision. The thigh from which the gracilis muscle has been dissected is now layed down in the adducted position, so that the gracilis muscle is relaxed (important). When the adduction is maximal, the ligament is pulled taut over a finger in the anus. The sutures are placed into the tendon and tied. The wounds are closed without drainage.

Postoperative management. The postoperative management is important. The patients must learn to defecate at specific times. Immediately before defecation is desired, the patient takes a laxative in the form of suppository, and sits on the toilet. In this position the gracilis muscle will be maximally relaxed and the anus will be in an open position. At the end of defecation the anus closes passively when the patient stands up. The patients cannot really be taught to contract and relax the muscle spontaneously.

Contraindication. Atonic rectum, diarrhea, severe constipation.

Results. These are reported as good provided that the patients are correctly selected, and that they are given thorough postoperative instruction and treatment. Only small series are reported in the literature.

Thiersch's ring

Indications. Rectal prolapse in high-risk elderly patients. This method should rarely be used because of its poor results and many complications. Delorme's operation or rectosigmoidectomy should be used instead.

Preoperative treatment. Bowel cleansing with Klyx®. Lithotomy position. Local or general anesthesia.

Technique (Fig. 16.5)
The steel wire or preferably a Marlex mesh can be placed subcutaneously around the anus in many different ways. The method described here is very simple:

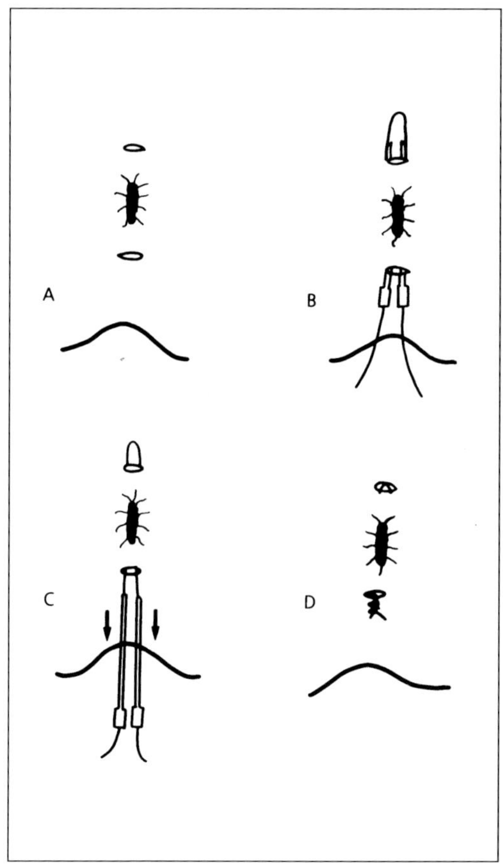

Fig. 16.5. *Thiersch's method.*

1. Two small incisions are made anterior and posterior to the anal canal.

2. Two large syringes are introduced via the posterior incision subcutaneously along each side of the anus and out through the anterior incision.

3. A steel wire is introduced through the posterior opening of one of the needles. It is then threaded through the anterior opening of the second needle until it emerges from the other end. The needles are now removed with the wire positioned subcutaneously around the anus.

4. The wire is now tightened over a finger in the anus by twisting the wire around itself with a forceps. When the wire is sufficiently tied the ends are removed with a wire cutter leaving the resultant knot beneath the skin of the posterior incision. The anterior and posterior incisions are closed. It is possibly more pleasant for the patient if a Marlex mesh is used.

Results. The results are poor with many complications in the form of infection, erosion of the wire through the skin, protrusion and constipation (fecal impaction). It is often necessary to remove the wire whereupon the prolapse recurs.

Delorme's operation

Indications. Old and senile patients with rectal prolapse.

Preoperative management. The bowel is cleansed as for the other operations in the region. Lithotomy position. The treatment is undertaken under general or local anesthesia.

Technique

1. The prolapse is evoked by pulling down the rectum with Allis's forceps. These are placed on the most distal end of the prolapse (Fig. 16.6).

2. Adrenaline solution (1:200,000) is injected submucosally allowing the dissection in the plane between the mucosa and the circular muscle layer to be undertaken more easily and ensuring hemostasis (Fig. 16.7).

3. A circumferential incision is made 1 cm proximally to the dentate line (Fig. 16.8).

4. The dissection in the plane between the mucosa and the circular muscle layer is made as far proximally as possible using a

cutting diathermy. The mucosal tube is excised (Fig. 16.9).

5. Hemostasis is secured with diathermy.

6. Plicating sutures are placed from the dentate line to the apex of the prolapse in the positions 12, 3, 6, and 9 o'clock and half-way between each of these (Fig. 16.10).

7. The sutures are tied crimping the prolapse like a concertina (Fig. 16.11). The residual mucosa is replaced in the anal canal by simple pressure (Fig. 16.12).

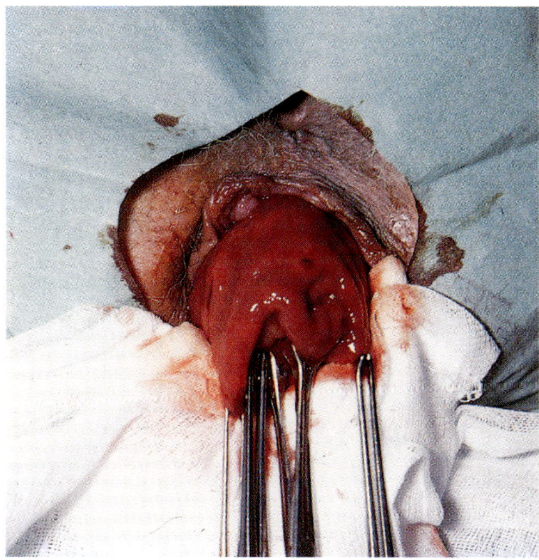

Fig. 16.6. *Delorme's operation. The prolapse is extruded through the anus.*

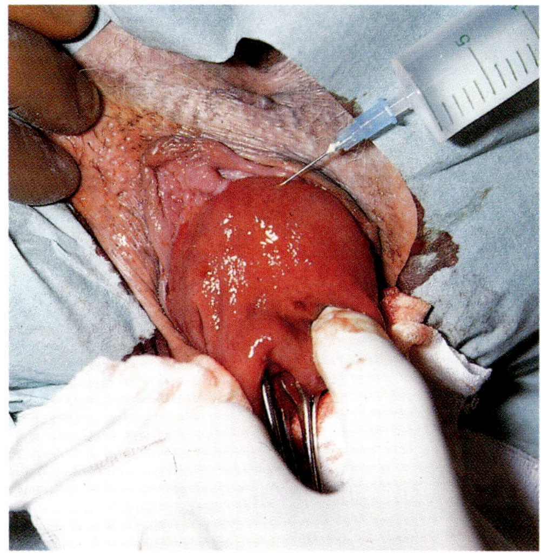

Fig. 16.7. *Separating the submucosa from the circular muscle by injection of lidocaine 0.5% with adrenaline.*

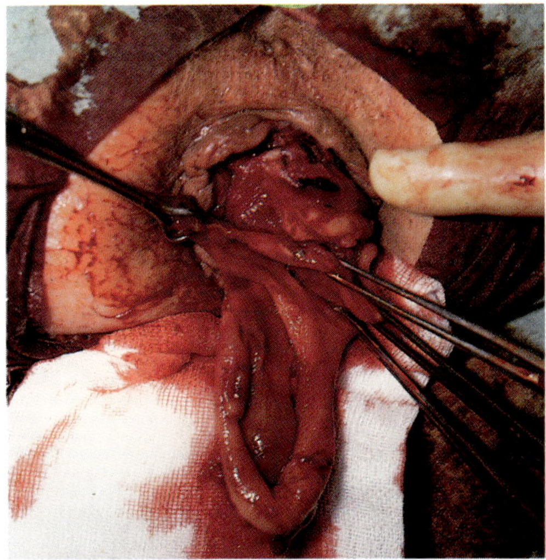

Fig. 16.8. *Mucosa is stripped from the underlying circular muscle after a circumferential incision of the mucosa 1 cm above the dentate line has been made.*

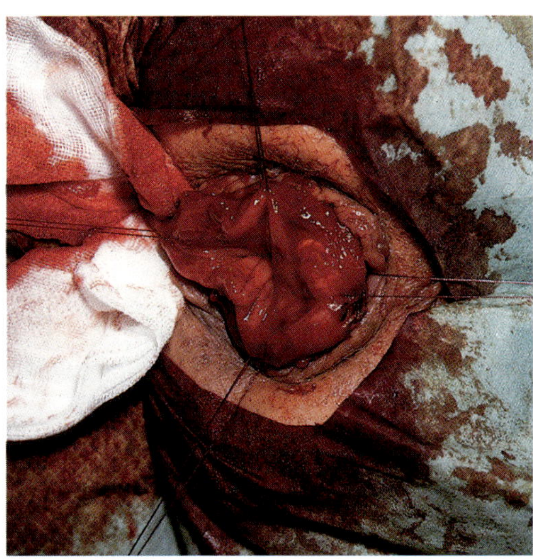

Fig. 16.10. *Denuded rectal wall is plicated with interrupted PDS sutures.*

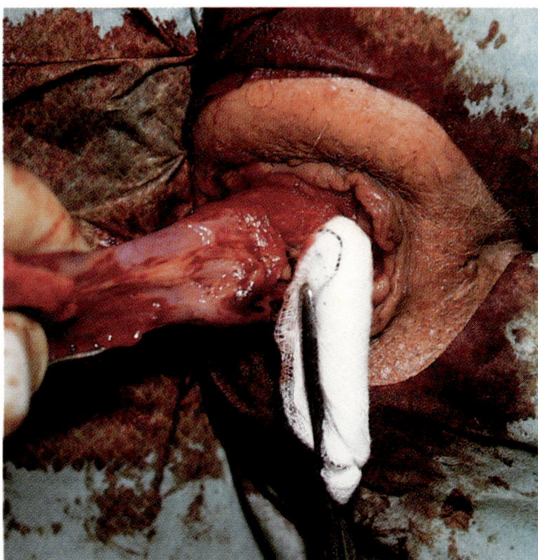

Fig. 16.9. *Mucosa has been stripped from the underlying muscle and is cut proximally.*

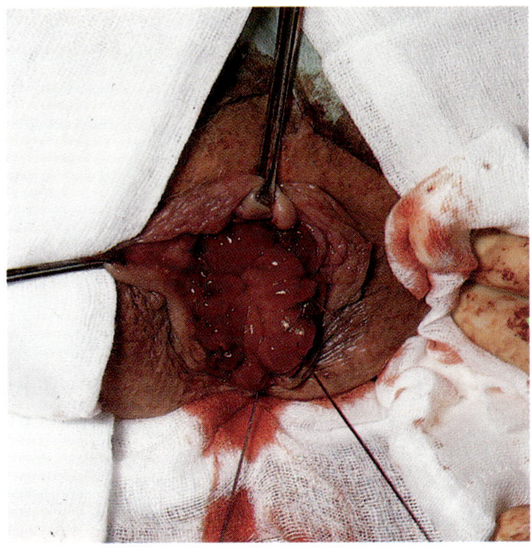

Fig. 16.11. *The sutures may be tied with the rectum prolapsed or within the rectum after it has been returned to the pelvic cavity.*

Postoperative management. Liquid diet until bowel movement. Bladder catheterization when necessary.

Results. The results after Delorme's operation are satisfactory in that the average recurrence rate is about 10% and varies in the literature from 0–24%.

Transanal rectosigmoidectomy

Indication. Rectal prolapse in old, senile, and infirm patients.

Preoperative management. Bowel cleansing. General, regional or local anesthetic. Lithotomy position.

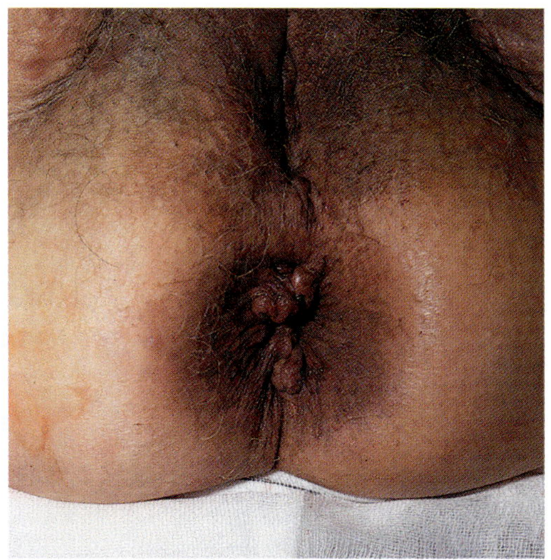

Fig. 16.12. *Delorme's operation: Final result.*

Technique (Fig. 16.13)

1. The prolapse is evoked by traction on the distal end of the prolapse with forceps (e.g., Allis or Babcock).

2. The rectum is incised circumferentially about two fingers above the anocutaneous junction and the mucosa and muscles layers are transsected. Hemostasis is ensured with diathermy or suturing.

3. The pouch of Douglas lying between the two layers of bowel is opened and the mesentery of the innermost bowel segment is identified posteriorly, and divided and ligated as high up as possible.

4. The opening into the pouch of Douglas is closed anteriorly with 3-0 sutures. The levator muscles are approximated anterior to the rectum with 2-0 sutures.

5. The outermost bowel segment is transsected, the rectosigmoideum pulled downwards, and the bowel transsected using straight Kocher's forceps 3 cm outside the anus.

6. After placing stay sutures in all corners an anastomosis is performed using interrupted 3-0 sutures. After the anastomosis is complete, it is pushed back into the anal canal.

Postoperative treatment. Bladder catheter. Liquid diet until bowel movement.

Results. The results after transanal rectal sigmoidectomy vary from series to series. The average recurrence rate in the literature is about 30%, and varies from 0–60%.

Artificial anal sphincter

Indication. Total anal incontinence caused by neuromuscular disorders such as myasthenia gravis.

Preoperative management. Total bowel irrigation. The patient is placed in the lithotomy position and the operation is undertaken under general anesthesia.

Technique

The technique is more or less the same as that used when positioning a prosthesis for urinary incontinence. The treatment should only be undertaken in departments specializing in proctology.

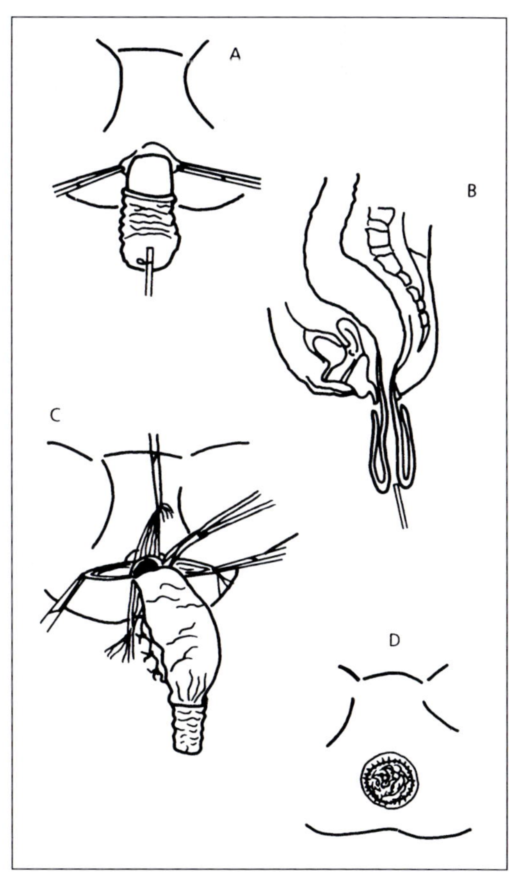

Fig. 16.13. *Transanal rectosigmoidectomy.*

1. Two vertical incisions of about 3–4 cm are placed anteriorly and posteriorly to the rectum (12 and 6 o'clock). Subcutaneous tunnels are created on each side of the anus via the incisions.

2. Anteriorly using the raphe of the transversus perineum and external anal sphincter muscles and posteriorly using the anococcygeal raphe, pockets are made so that the cuff is placed in the correct position around the anal canal.

3. Using subcutaneous tunnels the pump is placed in the left side of the scrotum and its pressure regulating balloon extraperitoneally to the left of the bladder.

Postoperative treatment. The patient is given antibiotic treatment during the operation and continued for the first three postoperative days. The cuff is kept empty for the first 10 postoperative days and the patient is given a liquid diet. Hereafter, the cuff is filled with air, and the patient learns to use the pump in the scrotum.

Results. The concept of this form of treatment for incontinence, resulting from neuromuscular disorders where operation such as muscle transplantation are not suitable, was developed at the Department of Surgery D, Copenhagen County Hospital at Glostrup, where the first prosthesis was implanted with success in a man with myasthenia gravis. The long-term results are unknown.

Colostomy

Indication. Senile, old people with incontinence with or without accompanying decubitus ulcer where other forms of continence operations are found to be contraindicated.

DESCENDING PERINEUM SYNDROME

Definition

A condition where the perineum descends 4–6 cm below the plane of the ischial tuberosities during straining.

Pathogenesis

The cause of this condition is not yet fully elucidated. Many of the patients present the same histological and electrophysiological abnormalities as are seen in connection with idiopathic anal incontinence. The abnormalities are most likely due to damage to the pudendal nerves or to the motor branches from the roots of S4. The nerve damage probably results from descent of the perineum of up to 4–6 cm during defecation whereby the terminal nerve branches are over-stretched causing irreversible damage to the nerves. Descent of the perineum may also be due to prolonged straining. The nerves may also be damaged during difficult and prolonged labor. It cannot, however, be excluded that a nerve lesion is the primary cause. Because of the resultant denervation atrophy of the muscles, the anorectal angle is straightened and the so-called flap valve mechanism is destroyed.

Symptomatology

The condition is most often accompanied by anal incontinence and defecation difficulties. Because of the destroyed flap valve mechanism, prolapse of the anterior rectal wall occurs into the anal canal. This mucosal prolapse may cause discharge or bleeding as is the case with prolapsed hemorrhoids. A number of patients complain of pain deep in the perineum, especially in the standing position. Characteristically the pain disappears on lying down.

Treatment

Treatment is directed towards avoiding exaggerated straining and constipation. The patients should be carefully advised about this and a diet rich in fiber is recommended. Mucosal prolapse can be treated by sclerotherapy, rubber band ligation or excision just as when treating hemorrhoids. If incontinence is present the patient may be offered postanal repair as described under anal incontinence.

Prognosis

The condition is difficult to treat. It recurs frequently. It is especially difficult to treat the associated perineal pain.

SUPPLEMENTARY READING

Browning GGP, Parks AG. Postanal repair for neuropathic faecal incontinence correlation of clinical results and anal canal pressures. Br J Surg 1983; 70: 101.

Cerulli MA, Nikoomanesh P, Schuster MM. Progress in biofeedback conditioning for fecal incontinence. Gastroenterology 1979; 76: 742.

Christiansen J, Kirkegaard P. Delorme's operation for complete rectal prolapse. Br J Surg 1981; 68: 537.

Christiansen J, Lorentsen M. Implantation of artificial sphincter for anal incontinence. Report of five cases. Dis Colon Rectum 1989; 32: 432.

Christiansen J, Pedersen IK. Traumatic anal incontinence. Results of surgical repair. Dis Colon Rectum 1987; 39: 189.

Corman ML. Management of fecal incontinence by gracilis muscle transposition. Dis Colon Rectum 1979; 22: 290.

Goldenberg DA, Hodges K, Hersh T, Jinich H. Biofeedback therapy for fecal incontinence. Am J Gastroenterol 1980; 74: 342.

Goligher JC. Surgery of the anus, rectum and colon. 5th ed. London: Baillière Tindall, 1984.

Parks AG. Post-anal perineorrhapy for rectal prolapse. Proc R Soc Med 1967; 60: 920.

Snooks SJ, Swash M, Setchell M, Henry MM. Injury to innervation of pelvic floor sphincter musculature in childbirth. Lancet 1984; ii: 546.

Stone HB. Plastic operation for anal incontinence. Arch Surg (Chicago) 1929; 18: 845.

Thiersch. Quoted by Carrasco AB. Contribution à l'etude du prolapsus du rectum. Paris: Masson, 1934.

XVII. Anal stenosis

Definition

An abnormal narrowing of the anal opening and/or anal canal.

Etiology

The stenosis may result from scar formation, pressure or invasion from an external process, a tumor or an inflammatory process in the anal opening or anal canal, which totally or partially occludes the lumen. The processes that cause stenosis may be malignant or benign. Benign stenoses are usually due to inflammatory reactions (Crohn's disease, ulcerative colitis, amoebiasis), trauma (often iatrogenic), radiation therapy or congenital abnormalities.

Inflammation

The stenoses of Crohn's disease are characterized by a transmural inflammatory process, scar formation, synchronous strictures in the colon and the severe perianal changes of Crohn's disease.

Trauma

The most common form of anal stenosis arises from trauma, such as hemorrhoid operations (fibrosis of the internal anal sphincter causing narrowing of the anus), recurrent and persistent trauma in connection with anal sex, low anterior resection of the rectum complicated by anastomotic leakage, the use of electrocautery in the region and local excision of malignant tumors (laser or conventional excision). The malignant tumors are most often primary malignant tumors of the anal canal or more rarely tumors of the surrounding structures, for example, the urinary tract.

Radiation damage

The earlier forms of radiation therapy for malignant tumors of the anal region often cause severe tissue necrosis with resultant scar and stricture formation. With modern radiation techniques symptomatic stricture formation is seen only in a few cases. The strictures are most frequently acompanied by radiation proctitis. The severity of the damage is dose dependent.

Chronic anal fissure and laxative abuse

Strictures may develop in relation to a chronic anal fissure. Chronic abuse of laxatives causes liquid stools, which do not dilate the anal canal, which gradually undergoes stenosis.

Sepsis

Sporadic cases of anal strictures resulting from sepsis are described in the literature. Patients with *Pseudomonas aeruginosa* infections and enterocolitis may develop ulcerating processes in the anal canal itself or perianally. These often heal with stricture formation.

Sexually transmitted anal diseases, such as lymphogranuloma venereum and gonorrhea, can cause the development of anal and/or rectal stenosis, although this is rare. The same is true for patients with tuberculosis and bilharziasis.

Classification

Strictures may be divided into diaphragmatic, annular or tubular. A diaphragmatic stricture is characterized by a thin band of tissue that closes the anal canal like a diaphragm. An annular stricture is less than 2 cm long. A tubular stricture is longer than 2 cm and is most often seen in the rectum.

Symptomatology

Most patients are asymptomatic. If symptomatic, patients complain of varying degrees of constipation, tenesmus, perineal pain, abdominal colic, rectal bleeding and ribbon-like stool. Patients with fibrosis of the internal anal sphincter after hemorrhoid operations have no or only mild symptoms.

Diagnosis

The diagnosis is made on physical examination. The etiological investigation should consist of X-ray examination of the colon. Rectoscopy and manual palpation under general anesthesia. Biopsy, examination of feces for ova and on suspicion of venereal disease tests for gonorrhea and lymphogranuloma venereum should be made. An anal profile investigation should be performed.

Treatment

The treatment depends on the degree and cause of the stenosis as well as its site. Short and low stenosis in the anal canal caused by chronic diarrhea can be treated with a fiber-rich diet. Strictures caused by hemorrhoids can often also be treated in this manner. Other conservative possibilities include the use of dilators such as Hegar's metal dilators or rubber dilators. Moderate or severe anal stenosis are best treated by anal plasty. The following methods will be discussed here: lateral subcutaneous internal sphincterotomy (by far the simplest method), Y-V plasty, S plasty, and quadrant sphincterotomy with or without advancement flaps.

Lateral subcutaneous internal sphincterotomy

Lateral subcutaneous internal sphincterotomy should only be used for patients in whom there is no loss of anoderm. Anal stenosis resulting from fibrosis of internal sphincter or hemorrhoid operations can be treated by lateral subcutaneous internal sphincterotomy if symptoms of the stenosis do not disappear spontaneously after dietary advice (see above). Anal stenosis associated with long-term chronic anal fissures or following prolonged chronic laxative abuse can also be helped by a lateral subcutaneous internal sphincterotomy.

Y-V plasty

Fig 17.1 demonstrates the principles of the procedure. This is used when there is significant loss of anoderm, for example, after too radical circumferential hemorrhoidectomy. An incision is made from the mucocutaneous junction and proximally into the anal canal where-

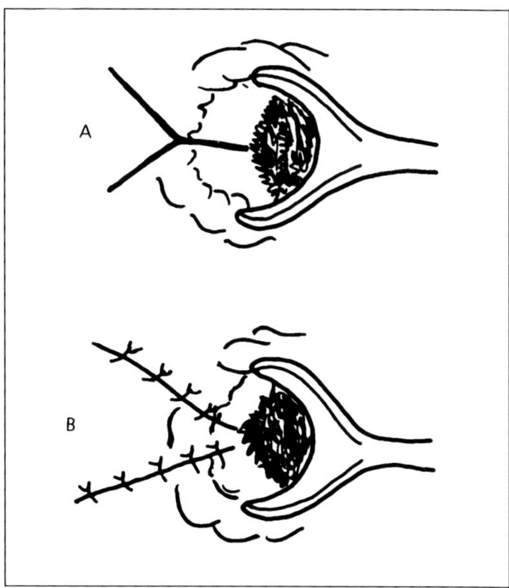

Fig. 17.1. *Y-V anoplasty.*

after the incision is extended in two directions towards the gluteal muscles, so that a Y-shaped incision results (Fig. 17.1 A). Full-thickness skin flaps are dissected which do not include subcutaneous tissue or the underlying muscle tissue. The base of the Y should be wide. The flap is undermined so that it can be advanced without tension. After immaculate hemostasis the apex of the skin flap is sutured to the base of the primary sphincterotomy incision (Fig. 17.1 B). Muscle and anoderm is sutured to each leg of the Y, so that it is transformed to a V (Fig. 17.1 B).

Note: Attention should be paid to the following: 1. The blood supply of the advancement flap. 2. Tension. 3. If the sutures are too tight, necrosis will result.

S plasty

S plasty is used for circumferential anal stenoses with loss of anoderm. A circular incision is made around the anus (Fig. 17.2) and all scar tissue is removed in a proximal direction. A partial sphincterotomy is carried out. With the anus as the center, an S-shaped incision is made so that a skin flap is developed on each side of the anus (Fig. 17.2). These are undermined up to their basis and rotated so that point b in Fig. 17.2 reaches point d, and point b* reaches point d*. After thorough hemostasis the

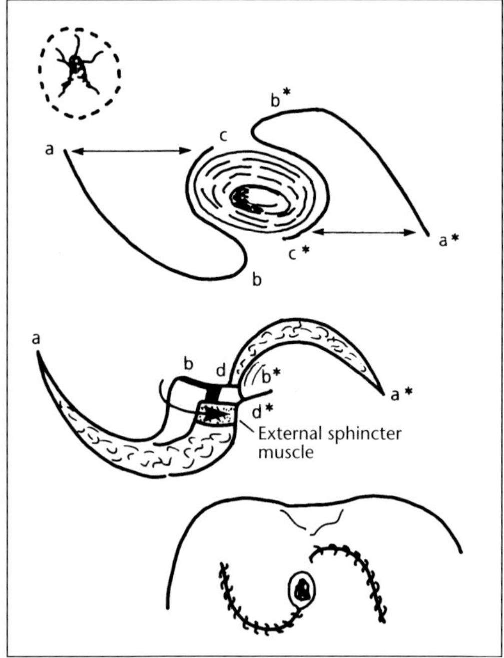

Fig. 17.2. *Principle of anal S-plastic.*

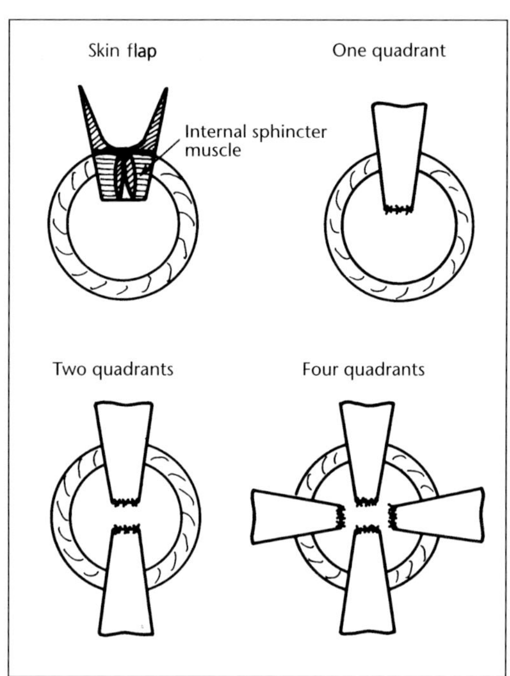

Fig. 17.3. *Quadrant sphincterotomy with or without advancement flaps.*

flaps are sutured so that a new mucocutaneous zone is created in the whole circumference of the anus (Fig. 17.2).

Quadrant sphincterotomy with or without advancement flaps

This is used for narrow anal openings where an ordinary sphincterotomy cannot be expected to increase the diameter of the anus. The technique is the same as for an ordinary advancement flap. Advancement flaps are made in two or four quadrants of the anus (Fig. 17.3).

It is true for all the advancement flap techniques that the well-defined procedure should be maintained in order to achieve good result. The rules are shown in Table 1.

Colostomy and/or excision of the rectum

It may be necessary in cases of severe anal stenosis resistant to other forms of treatment as, for example, with Crohn's disease or anal cancer, to employ a colostomy, or in cases or radio-resistant anal cancer to perform an excision of the rectum.

Table I. *Rules for the use of advancement flaps in the treatment of anal strictures*

1. Use always full thickness graft
2. Careful hemostasis is mandatory
3. Pre-, peri-, and postoperative antibiotics
4. Adjunctive internal sphincterotomy
5. Use absorbable sutures
6. Careful postoperative care and control
7. Patients with thin perianal skin should be excluded

SUPPLEMENTARY READING

Corman WL, Veidenheimer ML, Collier MA. Anoplasty for dual stricture. Surg Clin North Am 1976; 56: 727.

De Gosse JJ, Rhodes RS, Wentz WB, Reagan JW, Dworken HJ, Holden WD. The natural history and management of radiation induced injury of the gastrointestinal tract. Ann Surg 1969; 170: 369.

Ferguson JA. Repair of "Whitened deformity" of the anus. Surg Gynecol Obstet 1959; 108: 115.

Goligher JC. Surgery of the anus, rectum and colon, 5th ed. London: Baillère Tindall, 1984.

Greenstein AJ, Sachan DB, Kark AE. Strictures of the anorectum in Crohn's disease involving the anus. Ann Surg 1975; 181: 207.

Hudson AT. S-plasty repair of Whitehead deformity of the anus. Dis Colon Rectum 1967; 10: 57.

Kark AE, Epstein AE, Chapman DS. Nonmalignant anorectal strictures. Surg Gynecol Obstet 1959; 109: 333.

Ladd WE, Gross RE. Congenital malformations of the anus and rectum. Am J Surg 1934; 23: 167.

Nickell WB, Woodward ER. Advancement flaps for treatment of anal stricture. Arch Surg 1972; 104: 223.

Sarner SB. Plastic relief of anal stenosis. Dis Colon Rectum 1969; 12: 277.

XVIII. Spastic perineal syndrome

Definition

Constipation accompanied by defecation urge with frustated attempts at defecation and straining because of altered tone in the puborectalis muscle and the external anal sphincter. Consequently the patient is unable to extrude a 50 ml water-filled balloon from the rectum. This condition is also known as anismus.

Etiology

Unknown. It is likely that it is an acquired muscular dysfunction resulting from an involuntary reflex which causes muscle to contract or a voluntary suppression of normal relaxation.

Pathophysiology

As mentioned in Chapter 1 a normal defecation mechanism requires relaxation of the puborectalis muscle and the external anal sphincter. With spastic perineal syndrome the tone of the puborectallis muscle remains unaltered or displays a paradoxical dyssynergic contraction, when the patient strains. The condition can be compared with the well-known urological condition non-neurogenic detrusor sphincter dyssynergia or hyperactive urethra.

Symptomatology

In all patients the main symptom is a long history of constipation. The constipation is accompanied by a frustrated defecation urge. The majority of patients complain of bloating and abdominal pain of varying localization, intensity and duration. About 20–30% of the patients have pain localized to the left lower quadrant of the abdomen.

Diagnosis

The diagnosis should be suspected in patients with severe chronic constipation. The patient should be investigated by anorectoscopy, X-ray of the colon, and gynecological examination in females. In addition, the condition requires investigation in a proctological department with defecography, anorectal manometry, EMG, and assessment of colonic transit time.

Defecography will often show that 50–60% of the patients are unable to relax the puborectalis muscle, which is demonstrated by the fact that they have an unaltered or even a reduced anorectal angle while straining. Diagnosis is effectively made by placing a 50 ml water-filled balloon in the rectum and measuring the EMG activity in the puborectalis muscle (Fig. 18.1). All patients have an increased EMG activity. Fig. 18.2 shows the EMG and pressure relationships in a normal subject and in a patient with anismus. In about 50% of the patients the colon transit time will be prolonged with a terminal accumulation of the radioactive markers, which indicates that the pelvic floor is unable to relax (Fig. 18.3).

Treatment

As yet no adequate treatment is known. The management is often difficult. Many have attempted biofeedback treatment. Biofeedback treatment is done with the patient in the sitting position with a 50 ml balloon in the rectum to measure pressure and EMG activity. EMG activity can be made audible for the patient by a microphone enabling the patient to hear

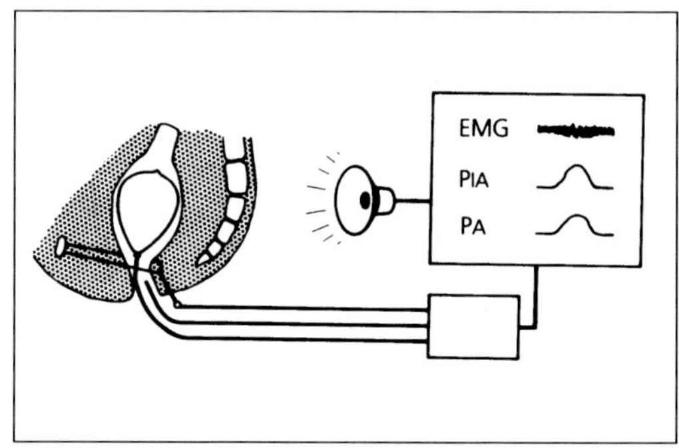

Fig. 18.1. *The principles of EMG and pressure measurements of the anorectum.*

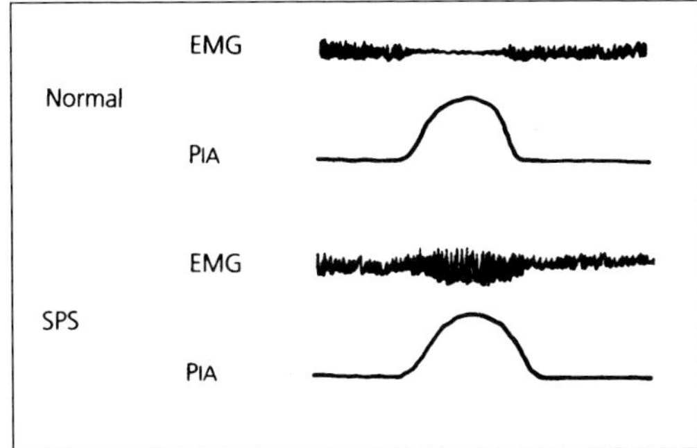

Fig. 18.2. *EMG and pressure fluctuations in a normal subject and in a patient with spastic perineal syndrome (SPS).*

whether the muscle is relaxed or contracted, and he or she attempts to extrude the balloon. In this way it is possible for the patient to relax the puborectalis muscle when straining. The patients are given a balloon to practice with at home between treatments, which last about 15 minutes and are given at 3–4 week intervals.

Prognosis

It is difficult to give a definite opinion on the effect of treatment. No controlled investigations exist. An extremely important prerequisite for the success of treatment is that the patient is motivated, otherwise treatment is useless. If the patient is well-motivated, it is thought possible to cure about 75% of the patients by the bio-

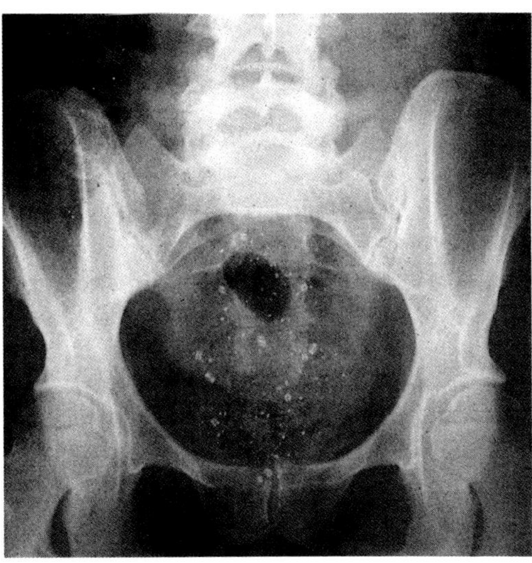

Fig. 18.3. *Increased colonic transit time in a patient with SPS with accumulation of radioactive markers in rectum.*

feedback method. If the patient has a large rectal capacity indicating reduced rectal sensation, the method fails even though the patient is taught to relax the puborectalis muscle while straining.

SUPPLEMENTARY READING

Buser WD, Miner PB Jr. Delayed rectal sensation with fecal incontinence. Successful treatment using anorectal manometry. Gastroenterology 1986; 91: 1186.

XIX. Anovaginal fistulae

Definition

A tube-like ulceration between the anal canal and vagina.

Epidemiology

Less than 5% of anorectal fistulae are ano- and/or recto-vaginal fistulae.

Classification

Ano- and recto-vaginal fistulae are divided into simple or complicated fistulae. The size of the fistulae varies. The majority are less than 1 cm in diameter.

Simple fistulae arise from trauma (iatrogenic) or infection, and are usually located in the lower part of the vagina. They are not especially large (less than 2 cm in length). They can be treated perianally without colostomy.

Complicated fistulae are larger (usually greater than 2 cm in length) and are usually located in the proximal half of the vagina. They arise as a result of radiation, inflammatory bowel disease or malignant tumors. They are treated by intraabdominal procedures with colostomy which is usually temporary.

Only ano-vaginal fistulae will be discussed here, that is to say low fistulae (Fig. 19.1), located just above the dentate line with a vaginal opening just inside the introitus of vagina. These ano-vaginal fistulae involve the sphincter muscle and to some degree the puborectali muscle.

Symptomatology

Some patients may be asymptomatic but most will complain of one or more of the following symptoms: unpleasant smelling discharge from the vagina, recurrent or chronic vaginitis or release of flatus and/or stool from the vagina. The severity of the symptoms depend on the size and location of the fistula as well as its course.

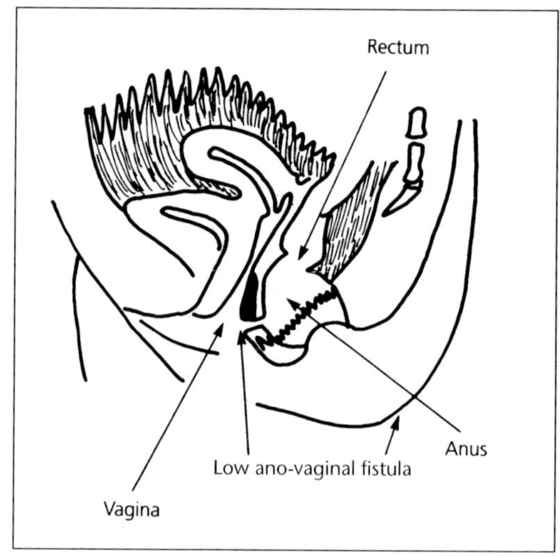

Fig. 19.1. *Schematic drawing of a low ano-vaginal fistula.*

Diagnosis

The diagnosis is made from a thorough history and physical examination which must include a gynecological examination and ano-rectoscopy. The patient should be investigated for the presence of Crohn's disease, malignant tumors, infection, and radiation damage. The fistula's size, exact location, and course must be described. Probing of the fistula can give an impression of its course. The function of the anal sphincter apparatus is examined as described in Chapter 1.

Etiology

Ano-vaginal fistulae may be congenital (see Chapter 4) or acquired. Only the acquired type will be discussed here.

Trauma

Iatrogenic damage of tissue during a gynecological procedure or obstetric trauma is thought to be the most common cause of ano-vaginal fistulae. If birth injury of the sphincter apparatus is not treated immediately, infection ensues in the wound, or if such lesions are inadequately managed, ano-vaginal fistulae or anal incontinence may easily arise. If a sphincteric lesion is treated immediately and correctly, good functional results are normally achieved.

Infective conditions

Infection in the ano-vaginal septum can cause an ano-vaginal fistulae. Ano-vaginal fistulae are described after perianal abscesses and different venereal diseases.

Specific infections

Specific infection of the ano-vaginal septa, such as Crohn's disease, ulcerative colitis, and tuberculosis, may result in an ano-vaginal fistulae.

Other causes

Malignant tumors of the vagina, anal canal or bladder as well as lymphomata, anaplastic anemia or agranulocytosis can cause ano-vaginal fistulae. It has been described that adenocarcinoma can develop in chronic ano-vaginal fistulae.

Treatment

A number of smaller ano-vaginal fistulae (less than 0.5 cm in length) heal spontaneously after iatrogenic and obstetric trauma. It is therefore advisable that these fistulae should not be closed operatively before 3–6 months have passed. Fistulae caused by inflammatory bowel disease almost never close spontaneously and the same applies to fistulae arising from radiation or malignant tumor.

Preoperative treatment

In general the better the patient's physical condition, the greater the chance of a good result of operative treatment of the fistula. It should therefore be ensured that the patient is in optimal physical condition preoperatively, possibly with the help of parenteral feeding and relevant medical treatment of the underlying disease. Adequate bowel cleansing is obligatory. The bladder is catheterized on the day of operation.

Operative techniques

There are many forms of surgical treatment for ano-vaginal fistulae. The method of treatment depends on whether the fistula is located above or below the puborectalis muscle. We will discuss only those procedures with which we have had personal experience.

Fistulae below the anorectal ring. If the fistula is located below the anorectal ring, treatment is simple and consists of excision of the fistula and suturing of the layers. The method often results in mild, or more rarely moderate, anal incontinence.

Anovaginal fistulae above the anorectal ring. Only two methods will be discussed, namely, conversion of the fistula to a complete perianal transsection followed by closure of the wound in layers (Fig. 19.2), or transanal covering of the fistula with an advancement flap of rectal mucosa (Fig. 19.3). The first operation is the same as that used by gynecologists for a third-degree vaginal tear where the anal sphincter and the rectal mucosa are involved. The steps of the operations are shown in Fig. 19.2. It is important that the walls of the rectum and vagina are dissected free from the sphincter, and that closure is done in layers. We use absorbable 3–0 interrupted sutures.

The best method for closure of an ano-vaginal fistula is thought to be the so-called advancement flap technique. It is best to position the patient prone (Fig. 19.3). The mucosa and submucosa of the rectum are incised down to the circular muscle layer (Fig. 19.3). Separation of the layers is easily done after injection of a solution of adrenalin 1:200,000 in saline into the area. Dissection should be done carefully to avoid tears in the flap. An adequate blood supply to the flap must be ensured. The base of the flap must be at least twice as wide as the apex to ensure the blood supply. When the flap has been mobilized adequately (at least 4–5 cm over the fistula opening) so that it can easily be advanced beyond the fistula. After the excess tissue containing the fistula opening has be excised, the flap is sutured firmly to the dentate line with absorbable 3–0 interrupted sutures (Fig. 19.3).

The result of the operation depends upon adequate mobilization of the flap, so that straining and retraction of the flap is avoided. The fistula opening on the vaginal side is left untouched. This means that blood and wound

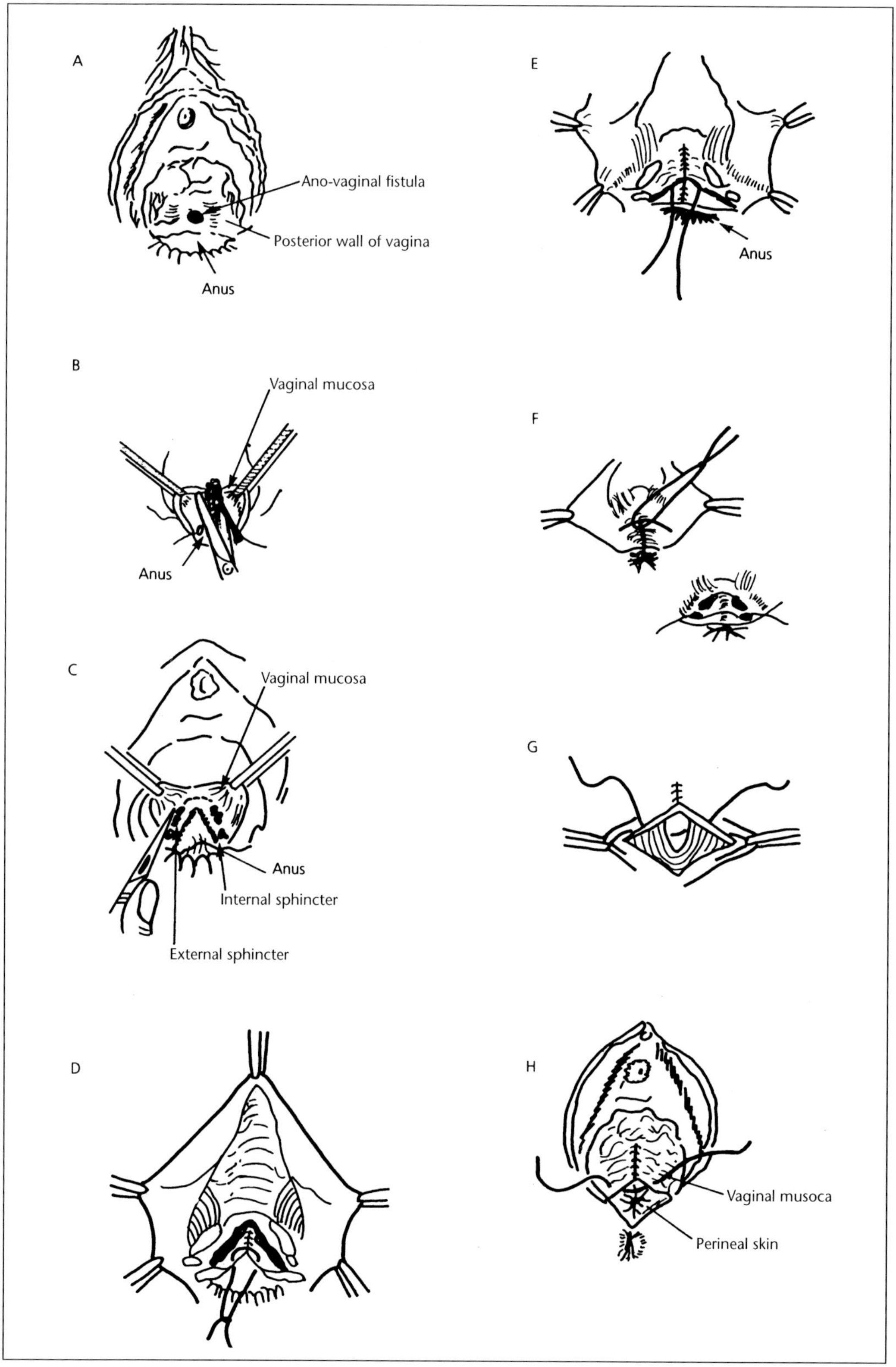

Fig. 19.2. *Complete transsection of ano-vaginal fistula and closure of the transsected layers.*

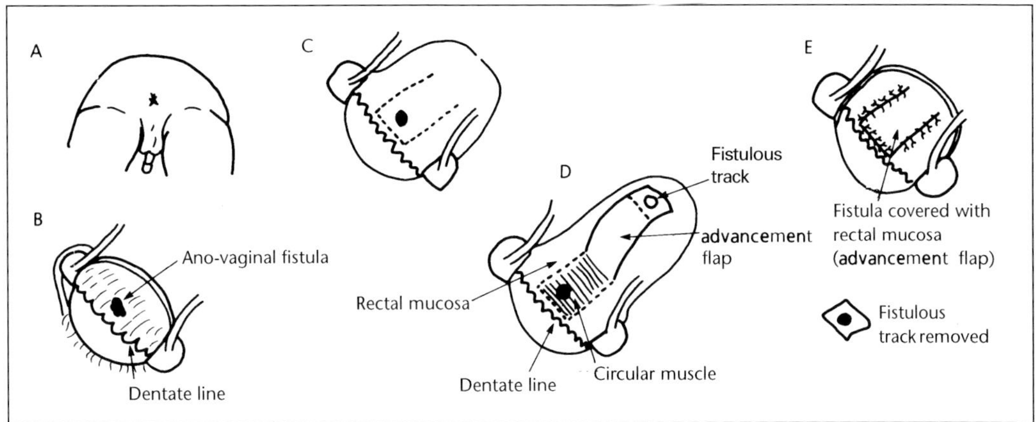

Fig. 19.3. *Endorectal advancement of rectal flap.*

secretion from the flap find a passage via the fistula to the vagina. Hemostasis is ensured by diathermy. One may choose to close the defect from the vaginal side at the same time using the advancement flap.

The results are considered good with this transanal advancement flap technique as about 90% of the fistulae remain closed after the operation.

Is colostomy necessary?

It is controversial whether it is necessary to employ a defunctioning colostomy before operating on an ano-vaginal fistula. It is our opinion that colostomy is not necessary when the ano-vaginal fistula has developed in normal tissue, for example after iatrogenic injury or obstetric trauma. On the other hand, we would never treat an ano-vaginal fistula arising in diseased tissue, for example, in Crohn's disease, radiation damage, tumors, or non-specific infection, without first having performed a defunctioning colostomy.

A permanent colostomy may be indicated in the treatment of radiation-induced high ano-vaginal fistulae, since one cannot use the radiation damaged tissue for local repair of the fistula because of severe tissue fibrosis. In some patients one can attempt transabdominal sphincter saving operations.

SUPPLEMENTARY READING

Corman ML. Anal incontinence following obstetrical injury. Dis Colon Rectum 1985; 28: 86.

Hoexter B, Labow SB, Moseson MD. Transanal rectovaginal fistula repair. Dis Colon Rectum 1985; 28: 572.

Jones IT, Fazio VW, Jagelman DG. The use of transanal rectal advancement flaps in the management of fistulas involving the anorectum. Dis Colon Rectum 1987; 30: 919.

Rothenberger DA, Goldberg SM. The management of rectovaginal fistulae. Surg Clin North Am 1983; 63: 61.

Russel TR, Gallagher DM. Low rectovaginal fistulas. Am J Surg 1977; 137: 13.

XX. Anesthesia for anal surgery

Many anal operations can be performed under local anesthesia. The reader is referred to Chapter 23 for details of the pharmacology, dosage, and choice of preparation of local anesthesia. Local anesthesia is suitable in operations for perianal hematoma, hemorrhoids, anal fissures, anal abscesses, condylomata accuminata, smaller tumors, and pilonidal cysts; however, it is less suitable for Milligan's operation for very large third- and fourth-degree hemorrhoids. Local anesthesia is unsuitable for anal fistulae and other larger procedures. In these cases as well as for especially nervous patients treatment under general anesthesia is recommended.

PREREQUISITES FOR THE USE OF LOCAL ANESTHESIA

When using local anesthesia whether it be during hospital admission or on an outpatient basis it is important that complete resuscitation equipment is always available. One should also be prepared for intubation to prevent aspiration, to give oxygen, and to administer fluid and various drugs intravenously. It is especially important not to use larger doses of local anesthesia than recommended.

PROBLEMS WITH THE USE OF LOCAL ANESTHESIA

The greatest problem when using local anesthesia in the anal region is the unpleasantness associated with the injection. It is important to prepare patients by informing them what is about to occur, and that the pain associated with the injection will be of short duration. Nervous patients may require preoperative sedation, such as 5–10 mg diazepam i.m. or i.v.

LOCAL INFILTRATION ANESTHESIA

This is the safest form of anesthesia with by far the fewest complications. A well-placed infiltration anesthesia gives short-term analgesia and relaxation of the sphincter muscles. When using large volumes the solution should contain adrenaline to counteract the vasodilatory effect of the anesthetic, thereby slowing absorption and reducing any toxic side-effects. It also minimizes capillary bleeding and prolongs the analgesic effect. On the other hand, the addition of adrenaline may mask delayed bleeding after the vasoconstriction has ceased. In anal operations done as outpatient procedures in which the patient is sent home immediately after the operation (less than 30 minutes after), adrenaline is not employed since it is better to observe any bleeding immediately and before the patient is on his way home.

Procedure

The perianal skin is first infiltrated circumferentially including the anal verge (Fig. 20.1 A). Then a finger is introduced into the anal canal and the anesthetic solution is infiltrated subdermally and submucosally all the way round (Fig. 20.1 B and C). With the finger in the anal canal one can clearly perceive the anesthetic solution as it is injected. The needle is introduced about 5–7 cm each time, and the injection is undertaken while slowly withdrawing the needle. Fluid should be injected anteriorly, posteriorly, and to both sides. Between 20 and 40 ml of 0.5% anesthetic solution is used. This infiltration anesthesia gives suitable analgesia and relaxation of the sphincter muscles so it is not necessary to inject local anesthetic directly into the actual sphincter muscle. Some, however, prefer to inject local anesthetic both submucosally as well as intramuscularly as shown in Fig. 20.1 (c).

In using infiltration anesthesia for perianal abscesses, the skin around and/or along the largest diameter of the abcess is infiltrated (Fig. 20.2). Fluid should not be injected

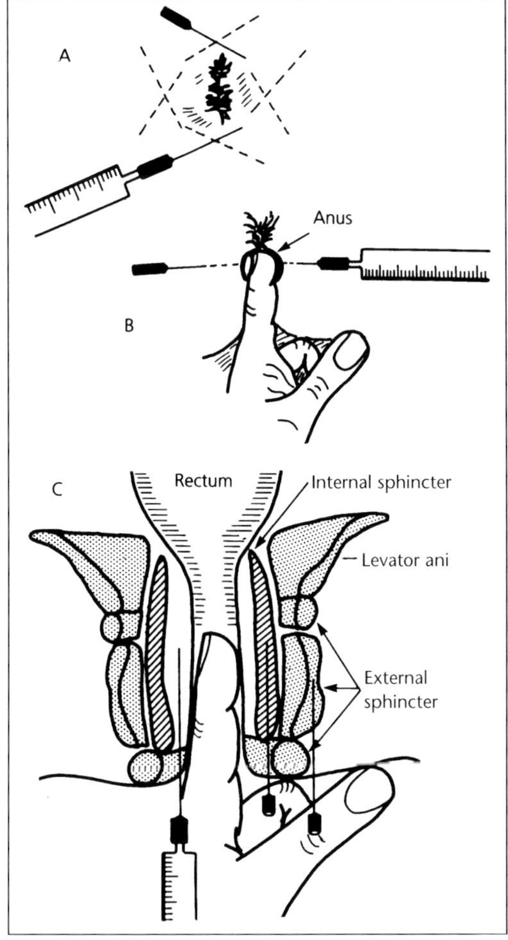

Fig. 20.1. *Anal surgery and infiltration anesthesia. A: Subcutanous infiltration; B: Intra-anal injection; C: Deep submucosal or intramuscular injection.*

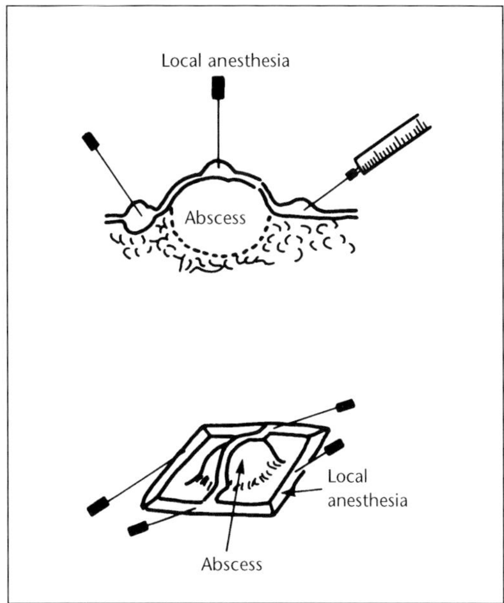

Fig. 20.2. *Infiltration anesthesia around and above an anal or a pilonidal abscess before incision.*

directly into the abscess cavity, since this will increase the pressure in the abscess, and cause the patient greater pain.

REGIONAL ANESTHESIA

Inferior hemorrhoidal nerve blockade is suitable for lateral subcutaneous internal sphincterotomy in the treatment of anal fissures. For this form of anesthesia the inferior hemorrhoidal nerve and the peripheral branches of the fourth and fifth sacral nerves are blocked resulting in paralysis of the external anal sphincter. This procedure should always be supplemented with local infiltration of the skin and subcutaneous tissue on the relevant side or circumferentially.

Procedure

The nerve blockade is done with the patient lying on the right side starting by infiltrating the skin 2–3 cm behind the anus. A finger is introduced into the rectum as a guide and deep injections of 10–15 ml of 2% lidocaine with adrenaline are performed on both sides of the anal canal. The effect occurs after 5–10 minutes.

LOW SPINAL ANESTHESIA

This form of anesthesia was regularly used in the past, but it is no longer recommended because of the relatively frequent complications. These are, first and foremost, postspinal headache in about 10% and urine retention in about 5% of cases. The method is further not recommended because of the very serious, although rare, complications of spinal nerve damage with subsequent incontinence for urine and feces, sexual dysfunction, or the most serious complication, paraplegia.

CAUDAL BLOCK (SACRAL ANESTHESIA)

This procedure involves injecting a suitable volume of local anesthetic into the sacral canal.

It gives good operative conditions with total analgesia if it works. There are significantly less complications combined with spinal blockade, but erroneous subarachnoid injection can result in a high block with total spinal anesthesia resulting in respiratory distress hypotension, and possibly coma. Unintentional intravenous injection can cause serious complications in the form of cramps, syncope and coma. Finally, it is a time-consuming form of anesthesia and often does not work because of erroneous placement of the needle. In some cases it is impossible to find the sacral canal either because of an inexperienced surgeon or because of a narrow sacral hiatus.

This form of anesthesia is only used when it can be performed by an experienced doctor.

Our preferred form of anesthesia for almost all cases is local infiltration anesthesia which is easy to employ and has few complications. Good analgesia and relaxation of the sphincter apparatus is achieved.

SUPPLEMENTARY READING

de Jong RH, Toxic effects of local anesthetic agents. JAMA 1978; 239: 1166.

Kratzer GL. Improved local anesthesia in anorectal surgery. Am J Surg 1974; 40: 609.

Moore DC. Regional block. 41th ed. Springfield: Charles C. Thomas, 1978: 19.

Ramalho LD, Salvati EP, Rubin RJ. Bupivacaine, a long acting local anesthetic in anorectal surgery. Dis Colon Rectum 1976; 19: 144.

XXI. Ambulatory anal surgery

Ambulatory treatment means that the patient is discharged on the same day or immediately after anal surgery. The treatment can be undertaken either in a private clinic or in a hospital out-patient department. In both situations facilities should be available to give the patient a general anesthesia if this becomes necessary. In this case the patient must remain under observation for most of the day. Patients who are treated under local anesthesia can, on the other hand, be discharged 30–60 minutes after completion of the surgery.

Contraindications

Patients with severe heart or kidney disease or other serious medical disorders which make the patient unable to look after themselves should in our opinion be treated as in-patients. Patients who are reluctant to undergo ambulatory surgery under local anesthesia should also be treated in hospital if the out-patient clinic does not have an anesthesiologist or safe recovery room facilities after general anesthesia. Patients with large third- or fourth-degree hemorrhoids, incarcerated hemorrhoids, large horseshoe abscesses, anal fistulae, or those requiring major procedures for tumors, strictures, or incontinence should obviously also be treated as in-patients under general anesthesia.

Advantages of ambulatory anal surgery

Patients with anal disorders are increasingly (>80%) being treated on an ambulatory basis in hospitals and especially in private clinics, where waiting lists are often shorter and where the patient is sure to be treated by the same specialist at each visit. Table 1 shows the anal disorders which can be treated with advantage as ambulatory procedures. The public health services can save a large amount of money if as many patients as possible with anal disorders are treated on an ambulatory basis. It is also more pleasant for the patient to obviate hospital admission. The patient can usually avoid general anesthesia and can be discharged immediately after the treatment and return to their familiar home environment. The decisive factor for a successful ambulatory treatment is without doubt meticulous patient counselling about the intended procedure, its possible complications and how the patient should deal with such complications should they arise.

Disadvantages of ambulatory anal surgery

Patients are largely left to themselves or possibly family members in the postoperative period. This absence of professional nursing care in the postoperative period can be compensated for by thorough preoperative counselling with both the patients and any relatives. A good relationship with the treating doctor is important. Patients treated on an ambulatory basis should be psychologically healthy. Many patients feel more secure with inpatient treatment, especially if they live alone. Patients should be informed how they can contact the treating doctor if any complications arise. Hygiene facilities (toilet and bathing facilities) at home should also be in order. It will often be helpful for a recently operated patient to receive daily visits by a nurse in the first few days after operation.

Patient counselling

The patient should feel secure about the forthcoming ambulatory operation and this requires thorough counselling. A well-informed patient is of great help to the surgeon. In the first place the treatment should be discussed thoroughly with the patient and the patient must be fully cognizant of the fact that the operation will be undertaken on an ambulatory basis. Written instructions concerning the type of operation, its possible complications and side-effects should be available so that the patient can go home and revise what the doctor has explained. The patient is often unable to take in the oral instruction because of nervousness. One should ensure that the patient understands the extent of the surgical procedure and how the local anesthesia will be given. The risk of postoperative complications and especially their prevention and treatment must be explained. The explanation should include the postoperative treatment as well as advice on the treatment of pain, regulation of bowel movement, general anal hygiene, and diet.

Treatment of pain

We never use medicines containing codeine but prescribe instead weaker analgesics about every 4 hours in the first postoperative day and thereafter as required. In some cases it may be necessary to prescribe or give stronger analgesics, for example, Morphine tablets six times a day for the first few days. We use this drug routinely after ambulatory hemorrhoidectomy and also recommend that out-patients take warm sitz baths two or three times daily lasting about 10–15 minutes each, especially after each defecation. Sitz baths have a relaxing effect on the sphincter muscles and are therefore pain relieving and also help keep the wound clean.

Regulation of bowel movement

A laxative such as psyllium (for example, Husk) twice daily together with an adequate fluid intake (at least 2 litres daily) regulate the motions and allow mild dilatation of the anal musculature which can have a favorable effect on anal pain arising from spasm. Another possibility is paraffin oil, 10–15 drops twice a day. This treatment has no dilatory effect on the anal muscles.

Anal hygiene

Dry toilet paper must not be used after defecation. Instead a sitz bath or a gentle hand-held shower should be used to clean the area. It is important that the region is dried thoroughly with soft cotton towel or a hairdryer.

Diet

The diet should be rich in roughage and this is best achieved by the intake of a large amount of plant fiber in the form of vegetables and fresh fruit. It is thought that the optimal daily amount of vegetable fiber is 35–40 g. The diet can be supplemented with a laxative, such as psyllium,

a few times a day together with adequate fluids (this should, however, be started after the patient has passed water postoperatively).

General guidelines following ambulatory anal surgery

Physical exertion should be avoided for the first 2 weeks after operation. Minor seepage of blood usually occurs, especially after defecation. Following rubber band ligation of hemorrhoids it should be explained to the patients that it is normal that some rectal bleeding will occur when the rubber band falls off 8–10 days later. If copious bleeding occurs the patient should contact the treating doctor and the patient should always be informed how he or she can seek help if the treating doctor is not available. If the patient is unable to pass water spontaneously after operation or is uncomfortable because of urine retention the treating doctor should be contacted. The same applies if signs of infection develop in the region of the operation. The patient should know that the first or subsequent defecations may be unpleasant because of pain. It is best that defecation occurs every second day. If the patient becomes constipated magnesium sulfate (10–15 ml) is prescribed. This can be repeated after 6 hours. If the constipation persists the patient should contact the treating doctor.

Follow-up

Patients should be followed-up once a week if there is a wound in the anal region and 3 weeks postoperatively a rectal examination should be performed to exclude the development of postoperative stenosis.

SUPPLEMENTARY READING

Jensen SL. A randomised trial of simple excision of nonspecific hypertrophied anal papillae versus expectant management in patients with chronic pruritus ani. Ann R Coll Surg Engl 1988; 70: 348.

Jensen SL, Lund F, Nielsen OV, Tange G. Lateral subcutaneous sphincterotomy versus anal dilatation in the treatment of fissure in ano in outpatients: a prospective randomized study. Br Med J 1984; 289: 528.

Jensen SL, Nielsen OV. Lord og Millar's børstemetode ved ambulant behandling af cystis pilonidalis. Ugeskr Læger 1988; 150: 98.

XXII. Complications after anal surgery

Complications after anal surgery are relatively rare and are, usually a result of poor surgical techniques. Some patient groups are more susceptible to postoperative complications, for example, patients with Crohn's disease, leukemia, diabetes mellitus as well as cancer patients and those undergoing immuno-suppressive treatment.

URINE RETENTION

Urine retention is the most common complication after anal surgery and requires catheterization in 5–10% of cases. It is due to the common innervation of the anal and urethral sphincters. Postoperative pain therefore causes reflex spasm of the urethral sphincter. This complication is twice as common in men as it is in women.

Preventative measures

Various preoperative measures can significantly reduce the risk of postoperative urine retention. The patient should urinate immediately before operation, preoperative intravenous fluid should be restricted and the patient's fluid intake in the first postoperative day should be kept to a minimum at least until the patient has urinated for the first time. Preoperative examination should reveal any prostatic hypertrophy since the risk of urine retention will be greater in such patients. Spinal anesthesia is more frequently complicated by urine retention than other forms of anesthesia.

Treatment

Treatment in the first instance is a cholinester (2–4 mg Carbacholine) if the patient becomes distressed and experiences pain and the bladder is distended and palpable over the symphysis then the bladder should be emptied with a catheter. If the patient is still unable to urinate a new catheter is inserted and left in place for about 2 days.

BLEEDING

Bleeding is especially common after hemorrhoid operations. As discussed earlier, bleeding may be early or late.

Early bleeding

This is almost always due to poor hemostasis and occurs after 1–2% of hemorrhoid operations. Diagnosis is easy and the patient should be readmitted to the hospital where bleeding can be stopped by suturing or rubber band ligation. Submucosal hematomas should be evacuated in order to prevent abscess formation.

Late bleeding

Late bleeding occurs about 7–10 days after surgery. There is often slight oozing in the first days after a hemorrhoid operation especially after defecation, but severe late bleeding may occur with sudden rapid loss of 0.5–1 liters of blood. This late bleeding is also seen after rubber band ligation. Late bleeding occurs in about 1–2% of cases either after straining during defecation with hard feces or spontaneously. The patient needs to be readmitted and may require a blood transfusion. The patient should be taken to the operating room and treated as described above.

INFECTION

Suppuration is an extremely rare occurrence after anal surgery. Infection in the anal region is thus seen in only 0.1–0.2% of hemorrhoidectomy patients. Patients with diabetes mellitus, Crohn's disease, leukemia or those under immunosuppression are, however, predisposed to postoperative infection. Serious necrotizing infection in the anal region can occur spontaneously but also occurs after anorectal surgery.

Symptoms

Severe pain together with fever and tachycardia are signs of infection in the wound. Crepitation and skin necrosis in the anal region are serious signs of delayed necrotizing infection.

Diagnosis

Thorough anal examination should be undertaken to reveal any cellulitis or abscess formation. Most abscesses arise in hematomas and these should therefore always be evacuated immediately.

Treatment

The treatment of abscess formation is described earlier. With necrotizing gangrene the treatment is broad-spectrum antibiotics, wide incision with débridement of necrotic tissue, and fashioning of an end colostomy.

Prognosis

There is a high mortality rate with necrotizing gangrenous infections of the anal region.

CONSTIPATION

Constipation or accumulation of hard feces in the rectal ampulla in the early postoperative period is caused by pain and fear of the first defecation after anal surgery. This condition can usually be prevented by employing a stool softener in the postoperative period. If the problem arises gentle water enemas are prescribed or oily laxatives are given to soften the feces. As a last resort, in rare cases, it is necessary to anesthetize the patient and digitally remove the hard fecal masses.

INCONTINENCE

Transient mild incontinence and soiling are sometimes seen after anal operations and are not due to lesions of the sphincter apparatus but to the fact that the patient soon after the operation has difficulty in distinguishing between the passage of flatus and liquid feces. This lack of sensory recognition is transient and lasts only a few weeks.

Persistent incontinence, which is a very serious complication, is seen almost exclusively after fistula surgery where large parts of the sphincter muscles have been transected. High incidences (10–20%) of incontinence after fistula operations are quoted in the literature. After lateral subcutaneous sphincterotomy continuous moderate or severe incontinence is almost never seen. Mild incontinence does occur. After hemorrhoidectomy transient or persistent incontinence of varying degrees are seen although they are extremely rare.

Preventative measures

To avoid these extremely unpleasant complications it is important to follow certain rules when performing operations for fistulae of the anal region. The puborectalis muscle should never be transsected; one should be conservative in treating fistulae which pass through the upper part of the external anal sphincter where the Seton method should be used. With lateral subcutaneuos sphincterotomy only the distal half of the internal anal sphincter should be transsected. Persistent incontinence is seen in about 5% of patients after posterior fissurectomy; for this reason the method should not be used. In order to prevent problems with continence after hemorrhoidectomy the sphincter musculature should not be included when ligating the hemorrhoid pedicel.

Treatment

see Chapter 16.

ANAL STRICTURE

Anal stricture formation after anorectal surgery is a rare and relatively late complication. The incidence is not known but it probably lies between 1 and 3%. The stricture may be a progressive narrowing of the anal verge and in many cases is due to excessive removal of anoderm and/or perianal skin. Exaggerated use of electrocoagulation or infections in the anal region can cause stricture formation. Fibrosis and scar tissue gradually narrow the anal canal.

Preventative measures

It is important to preserve sufficient anoderm and skin during hemorrhoidectomy (preferably leave a single hemorrhoid behind) and to restrict the use of electrocoagulation when treating condylomata acuminata (these should rather be removed by simple surgical excision). Daily dilatation of the anal canal after hemorrhoidectomy, for example, is used by some but is not recommended since recurrent trauma of the wound eventually results in a stricturing scar tissue.

Treatment

See Chapter 17.

DELAYED WOUND HEALING

Delayed wound healing is often due to excessive removal of anoderm or daily instrumental dilatation of the anal canal. Overhanging wound edges (insufficient trimming) can grow together before the actual wound has healed and may cause undrained pockets with possible infection discharge and pruritus ani. If the wounds have not healed after 10–12 weeks the patient should be reexamined to exclude Crohn's disease or an undiagnosed cancer. Granulation tissue should be curetted and treated with 10% silver nitrate solution. If pocket formation has occurred débridement should be undertaken.

ADDITIONAL READING

Bennett RC, Friedman MHW, Goligher JC. Late results of haemorrhoidectomy by ligature and excision. Br Med J 1963; 2: 216.

Campbell ED. Prevention of urinary retention after anorectal operations. Dis Colon Rectum 1972; 15: 69.

Goligher JC. Surgery of the anus, rectum and colon, 5th ed. London: Ballière Tindall, 1984.

Marks CG, Ritchie JK. Anal fistulas at St. Mark's Hospital. Br J Surg 1977; 64: 84.

Watts JM, Bennett RG, Dutchie HL, Goligher JC. Healing and pain after haemorrhoidectomy. Br J Surg 1964; 51: 808.

XXIII. Anal pharmacology

The majority of drugs that are used in the treatment of anal disorders are administered through the skin or via the rectum. Oral medicines are generally only used for infectious diseases, most often sexually transmitted.

Since anal disorders are common in the population the use of medicines for these disorders is extensive. There are many locally active medicines which do not require prescription and even though their effect is extremely doubtful they are commonly used. The same may be said of many of the prescribed preparations.

CUTANEOUS

Skin preparations are used locally in the treatment of many anal disorders where systemic absorption is not intended. The purpose of local treatment is to achieve a high local concentration of the preparation without incurring systemic side-effects. With certain skin diseases of the anal region, however, the risk of systemic absorption is increased (eczema, maceration) since the barrier function is diminished. This should be recognized in using preparations such as steroid ointments and podophyllin.

Ointments and creams

The preparations used locally in the anal region are ointments, creams and various semi-solid or fluid medications. Many of these preparations are prescribed for use within the anal canal. If this is the case it is important to explain to the patient that it does not help to apply the ointment or cream to the anal verge. The preparation must be introduced into the anal canal either with a gloved finger or a special applicator which comes with the preparation. If an effect is desired in the anal canal a suppository should be prescribed instead since many patients regard the introduction of ointment into the anal canal as unesthetic. In order to protect the hands against the preparation the patient should be instructed to use gloves when applying the ointment.

The disadvantage of this method of application is the risk of developing local allergic reactions or skin irritation.

Suppositories

Suppositories are often used in the treatment of hemorrhoids and anal fissures. Suppositories usually consist of a mixture of medication and a carrier (for example, cocoa butter) which melts at body temperature. Suppositories should therefore be stored at low temperatures. The patient is instructed to introduce the suppository with a gloved hand and wash the hands afterwards.

The disadvantage of suppositories is that the rectal mucosa is easily irritated by the active ingredient.

Tar preparations

In an attempt to avoid the use of local steroid preparations for anal disorders we have attempted to use tar-containing skin preparations in the treatment of pruritus ani since these preparations have an antiseptic and antipruritus effect.

Indications

Idiopathic and secondary pruritus ani.

Method of action

The method of action is not known.

Side-effects

Contact dermatitis, allergy, and photosensitivity in the case of some preparations.

Choice of preparation and dosage

Many different preparations may be used. We have only tried tumenolammonium, a water-soluble slated tar, that does not stain, have a unpleasant smell or cause photosensitivity. Earlier we have used a cream containing

600 mg tumenolammonium, 2.4 g olive oil, and 27 g lanolin. Since lanolin (animal fat) is liable to cause allergy reactions we now use a commercial preparation 2% Tumedan (Danapharm, Espergærde). The preparation is applied to the perianal skin morning and evening as well as after each bowel movement.

Corticosteroids

Glucocorticoid-containing preparations are often used in the symptomatic local treatment of many anal disorders, especially hemorrhoids, anal fissures, and pruritus ani. For many of these indications the clinical effect of the steroid is doubtful and controlled randomized studies are lacking. Many different preparations are available containing either pure glucocorticoids or steroids in combination with antibiotics and local anesthetics. No differences have been shown in the clinical effect of the various preparations. Glucocorticoid-containing preparations for local use may be divided into four groups according to their strength. We use only preparations from group 1 (weak glucocorticoids).

Method of action

Corticosteroids act mainly by suppressing the inflammatory processes regardless of the etiology (infection, physical or chemical irritation). Steroids inhibit the release of the biological substances which provoke the inflammatory response (histamine, prostaglandines, bradykinin, leukotrines). They also have an effect on blood vessels. Glucocorticoids do not cure the diseases but they relieve the symptoms for as long as the treatment is employed. Glucocorticoid treatment should be used only for non-infectious conditions since they can exacerbate infectious conditions.

Side-effects

The use of anal medications containing glucocorticoids can cause both local and systemic side-effects. The latter is, however, rare and mostly of theoretical interest since glucocorticoid absorption in the anal region is extremely limited with the preparations we normally use.

The local side-effects occur after either short- or long-term use of the preparations. After long-term treatment especially with a strong glucocorticoid preparation (rectal Betnovat) so-called "steroid skin" may arise. This is caused by atrophy of the subcutaneous connective tissue. The skin becomes thin and transparent. Superinfections or contact dermatitis may arise even after a few days of treatment with pure steroid ointment or cream. We have seen severe herpes analis infection occurring after a few days treatment with weak steroid cream (2% hydrocortisone) in two patients. A so-called rebound phenomenon can occur on stopping treatment with a strong steroid preparation. This presents with redness and a burning sensation in the perianal skin. The symptoms disappear after 1–2 weeks. This can be avoided if the treatment is gradually reduced by shifting to a weaker steroid preparation.

Contraindication

Infectious anal disorders. The preparations may be used in pregnant or nursing patients.

Choice of preparation and dosage

We rarely use local steroid treatment of anal disorders and if we do we choose a weak pure steroid preparation since preparations containing antibiotics and/or anesthetics are more likely to cause allergic reactions and contact dermatitis. Side-effects are extremely rare when using weak steroid preparations (1–2% hydrocortisone) as opposed to the use of strong preparations for example betamethasone (rectal Beetnovat). Hydrocortisone is available as an ointment or cream in concentrations of 1–2%. Proctosedyl® is an example of a combination preparation which is commonly used in Denmark in the local treatment of various anal disorders. The preparation is available as suppositories and as an ointment. It is an anal preparation containing a surface-acting analgesic of the amide type, a glucorticoid, and an antibiotic.

It is emphasized that if the ointment is to have an effect in the anal canal it is insufficient to merely apply it to the perianal skin. It must be introduced into the anal canal.

Cleansing agents

In the treatment of idiopathic and secondary pruritus ani it is often best to advise the patients to avoid the use of soap and water after bowel movements since they sometimes cause irritation and skin dryness. Softening and protecting

Table I. Contents of Bidet$^{(R)}$* emulsion*

Non-ionic emulgation
Liquid paraffin
Isopropylmyristate
PCL-liquid
White vaseline
Di-sodium-monolauramido-sulfosuccinate
H_2O (deminiralised)
Phenonip (preservative)

cleansing agents should be used instead. Many different agents are available (silicone preparations, zinc oxide preparations, vaseline, fatty materials and carbamide preparations). We will discuss only the skin cleansing agent Bidet® since we have no experience with other agents in the treatment of disorders of the anal region.

The composition of Bidet® cleansing emulsion is given in Table 1. The emulsion contains neither color nor perfume and is available in bottles of 60 ml.

Indications

Idiopathic and secondary pruritis ani and nappy rash in children as well as all irritative conditions of the anal region.

Dosage

It is applied morning and evening as well as after each bowel movement in a thin layer with a soft cloth, cotton wool or similar material. After cleansing the perianal skin is dried with a soft cloth.

Contraindications

Allergy to the components of the emulsion. We have not encountered side-effets amongst 200 patients treated with this emulsion.

VEGETABLE FIBER AND BULK-FORMING LAXATIVES

The diet of the Western world may be characterized amongst other things by its higher sugar content and lower content of roughage or vegetable fiber than that found in underdeveloped countries. The lack of vegetable fiber in the diet of the industralized world has been shown to be of importance in the development of many anorectal disorders, such as hemorrhoids, anal fissures, and idiopathic pruritis ani. Danish investigations have shown that patients with these disorders have a lower intake of vegetable fiber than healthy subjects. Moreover, it has been shown that a supplement of vegetable fiber in the form of wheat bran reduces the recurrence rate of both hemorrhoids and anal fissures. There is therefore an increased tendency to recommend a low sugar diet with a greater fiber content or to prescribe fiber supplements to the normal diet of these patients. This can be achieved by using bulk-producing products such as wholewheat bread or other bran-containing bread as well as vegetables and fruit. If the patient is reluctant to institute such dietary changes or cannot afford it they can be advised to use ordinary wheat bran or a bulk-producing laxative such as the pulverized husks of white cilium.

The agents require no prescription in Denmark and are available in pharmacies, health food stores, and supermarkets.

Wheat fiber

Wheat fiber is a byproduct of flour production. In the production of wheat flour the indigestable husks are separated in the form of bran. This is an extremely cheap product which may be given as a dietary supplement whereby the symptoms of disorders such hemorrhoids, anal fissures, and colonic diverticulosis may be reduced or totally abolished.

Method of action

Bran decreases the intraluminal pressure in the anorectal region without affecting the basic motility of the colon and rectum. Another important feature of fiber is that it can absorb four times as much water than its own weight, i.e., 100 g of fiber binds more than 400 g of water. The result of this is increased bulk of the feces. Bran and other vegetable fibers have a greater effect on the volume of stool than protein, fat or carbohydrate residues. The increased volume of feces decreases the transit time in the colon and results in mild physiological dilatation of the anal canal and its sphincter apparatus. This is a possible explanation of bran's favorable effect on the symptoms of hemorrhoids and fissures.

The effect on the weight of the feces also depends on the individual's sensitivity to vegetable fiber and the presence of bacteria in the feces. Much of the water content of feces arises from bacteria and bran and other vegetable fibers promote the growth of bacteria in the colon. The patient should be instructed that the laxative effect may be delayed, that is to say, a few days and as long as a few weeks.

Dosage

Ordinary wheat fiber is prescribed in doses of 5 g morning and evening in the first instance and is increased to 10 g morning and evening thereafter.

It is important to instruct the patient to take the bran mixed with water or other liquid (about 200 ml of liquid per dose) and, moreover, to drink copiously (at least 2 l of liquid per day).

Side-effects

Almost all patients that are treated with bulk-forming laxatives complain of bloatedness during the first weeks. This effect is usually transient and should be explained to the patient who should not stop the treatment for this reason. If the patient does not drink sufficient fluid during the treatment with bran or a similar agent dry lumps of fiber are formed which may obstruct the esophagus or small bowel. Many patients find it unpleasant to take bran mixed in water. It has no taste and it is unpleasantly sticky in the mouth. It is, however, by far the cheapest bulk-forming laxative.

Psyllium

Pulverized husks of psyllium seeds. This is a softening and bulk-forming laxative similar to bran. Depending on the type it can swell to 20–40 times its volume when mixed with water. Many different types are available. We use either Husk or Vi-Siblin. The advantage of these subtances over bran is that they are much more pleasant to ingest. Husk is the cheaper of the two.

Dosage

5–10 ml morning and evening is normally recommended together with a large volume of liquid as mentioned in connection with wheat bran.

Side-effects

These are similar to those encounted with wheat bran. Ingestion of very large amounts can cause bowel obstruction.

LOCAL ANESTHETICS

Local anesthetic agents may be divided into two groups. The first consists of a series of esters and the other of amides. Local anesthetic agents of the ester group include, amongst others, benzocaine, procaine, and cocaine while the amide group contains, amongst others, lidocaine and mepivacaine. We tend to use either lidocaine or mepivacaine with or without the addition of a vasoconstrictor.

Mode of action and method of use

The agents work by blocking the transmission of impulses along nerves and across nerve endings. The effect is reversible. The efficacy of the agents depends on their concentration at the nerve fibers. The latent period before the agent works depends on the preparation and method of administration. Lidocaine works more quickly than mepivacaine. The function of the preparation is increased by the addition of a vasoconstrictor for example adrenaline or noradrenaline concentration of 1:200,000 and this allows a higher concentration to be used. Anesthesia is fully effective after 2–5 minutes and lasts aboout 1–2 hours or up to 4 hours if a vasoconstrictor has been added. If vasoconstrictor is contraindicated mepivacaine should be used instead of lidocaine.

Local anesthetic agents are also used topically in the anal canal and perinanally as surface analgesics. The agents are used for infiltration anesthesia for ambulatory surgical treatment of anal disorders.

Complications

Vaso-vagal attack. This may occur in anxious patients and in patients with severe pain. It is seen while the local anesthetic is infiltrated. Vaso-vagal attacks occur in about one out of 1000 patients receiving infiltration anesthesia of anal canal.

The symptoms are nausea, sweating, pallidity, and temporary unconsciousness which may be accompanied by clonic convulsions, bradycardia, and hypotension (Table 2).

Treatment consists of placing the patient in Trendelenburg's position, administration of oxygen, and intravenous atropine (1 mg) if the bradycardia does not disappear spontaneously.

Table II. *Complications after local anesthesia*

Vaso-vagal reactions
Systemic reactions
  Paresthesias of tongue and mouth region
  Muscle fasciculations
  Blurred vision
  Confusion
  Convulsions
  Coma
  Cardiac arrest
Allergy

Systemic reactions. When using very high concentrations systemic reaction may occur. For this reason local anesthetic agents containing vasoconstrictor should be used if it is necessary to use large amounts of anesthetic. Intoxication develops 15–30 minutes after administering infiltration anesthesia. The symptoms depend on the concentration of the anesthetic agent in the blood and include paresthesia of the tongue and oral region, confusion, visual disturbance, muscle fasciculations, unconsciousness, spasms, circulatory failure, and cardiac arrest. Hemodynamic changes occur only with very high concentrations of the agents.

Treatment consists of manual ventilation with oxygen through a face mask. In the event of spasms diazepam or thiomebumalnatrium in doses of 10 and 100–200 mg, respectively, are given. Most important though is manula ventilation and administration of oxygen.

Allergy occurs but this is very rare and includes skin rashes, Quincke's edema, and possibly anaphylactic shock. It is treated with antihistamine, adrenaline or steroids.

Dosage

The maximal dosage for lidocaine is 200 mg and 500 mg with vasoconstrictor. The corresponding dosages for mepivacaine are 300 and 500 mg, respectively.

Preparations

Lidocaine (Leostesin® Lidocaine®, Xylocaine®) made in 0.5, 1, and 2% solutions with and without vasoconstrictor for use as infiltration anesthetics. They are also available as ointments (Leostesin®) and suppositories (Lidocaine®) for surface analgesia.

Mepivacaine (Carbocaine®) is available in solutions and concentrations mentioned above but without the addition of vasoconstrictors.

CYTOSTATICS
Podophyllin resin

This is made from the dried roots of the plant *Podophyllum peltatum*. It is an amorphic powder of various colors with a characteristic smell and bitter taste. It is poorly soluble in water but is totally soluble in alcohol. In Denmark it is used only in the treatment of condylomata and warts.

Method of action

It has no specific effect on the virus but is a specific inhibitor of mitosis. The substance is absorbed through the skin or mucosal membrane and can cause systemic effects.

Dosage

It is applied to condylomata once a week until they disappear. Not more than a total of 1 ml should be applied at a time because of the risk of systemic side-effects. It should be applied only to the condyloma itself and not onto normal skin or mucosal membrane.

Complication

In our opinion treatment should be done by or under supervision of a doctor. About 4–8 hours after application the podophyllin is washed away.

It should not be used to treat small children or pregnant women even though its use has been reported in 90 pregnant women without effects on the fetuses.

Side-effects

Local side-effects are seen if excessive amounts are used and if the substance is applied to the surrounding normal skin and if the patient does not remember to wash the area. Severe irritative changes may be seen in the skin especially if superinfection occurs. Thrombocytopenia and leucopenia as well as lethargy and respiratory problems have been seen with local application of podophyllin. Neurotoxic complications have also been described in the form of polyneuropathy after application of about 2 g of podophyllin.

Preparations

Liniment 25%: Podophyllum in liquid paraffin (10 ml). Liniment containing 20% alcohol (podophyllum in spiritus fortis) (10 ml). Podophyllotoxin: Cream 0.15% or liniment 5 mg/ml.

5-fluorouracil (5-FU)

A synthetic antimetabolite analog to the naturally occurring precursor in the purine and pyrimidine synthesis chain.

Indications

Parenteral: Squamous cell carcinoma in the anal canal in combination with mitomycin.

Topical: Bowen's perianal disease and with precancerous lesions of the perianal region.

   Parenteral administration should only be undertaken in conjunction with an oncologist.

Dosage

Parenteral: Intravenous administration of 10–15 mg per kg body weight once a week. It is given in combination with mitomycin.

Topical: A cream is applied once to twice daily in a thin layer without dressings for about 6 weeks.

Side-effects

Parenteral: Nausea, vomiting, and bone marrow depression.

Topical: Local redness and a burning sensation which disappears with local application of steroid cream or ointment without affecting the action of the antimetabolite.

Contraindications

Lactation, pregnancy, allergy, and bone marrow depression.

Preparations

Injection fluids: 50 mg/ml ampules of 5 ml. Cream 5% in tubes of 20 g.

Mitomycin C

An antineoplastic antibiotic produced by *Streptomyces caespitosus*. Violet-blue crystalline powder, soluble in water.

Method of action

The substance is alkylating and depresses the synthesis of nucleic acids. It is a highly toxic antineoplastic agent that especially in other countries is used in combination with other agents in the treatment of gastrointestinal tumors. In the United States the substance has been used in combination with 5-FU in the treatment of invasive squamous cell carcinoma of the anus with good results.

Administration and dosage

The substance is given intravenously for anal cancer but can also be given intraarterially. For squamous cell carcinoma of the anus it is given only as a single-dose intravenous bolus on the first day of treatment (15 mg/m^2) followed by 5-FU treatment.

Side-effects

Cumulative bone marrow depression may be seen. This is usually reversible about 4 weeks after ceasing treatment. Other serious side-effects are kidney and lung damage. The drug should therefore not be used in patients with poor kidney function. Necrosis occurs if the agent is injected paravenously.

Preparations

Injection fluid in ampoules.

Bleomycin

This is a antineoplastic mixture of glycopeptide antibiotic containing bleomycin A$_2$ and B$_2$ which is produced from the growth of *Streptomyces verticillus*. Cream-colored powder which is highly soluble in water.

Method of action

The drug bind itself to the cells' DNA and causes rupture of the DNA chains. It is used for many different tumors. In the anus it can be used to treat squamous cell carcinoma both in the perianal region and the anal canal. There is no-one in Denmark that has broad experience with this drug in the treatment of squamous cell carcinoma in the anus.

Administration and dosage

The drug can be administered subcutaneously, intramuscularly, intravenously or intraarterially. The usual dosage for squamous cell carcinoma is 15 to 30 mg per week in divided doses. Continued infusion of 15 mg per day for up to 10 days is used for squamous cell carcinoma of the skin and anus.

Side-effects

The commonest side-effects are skin rash, itching, blistering, hyperkeratosis, nail changes, and stomatitis. Mild marrow depression can occur, thrombophlebitis, and fever are seen after parenteral administration. The most serious side-effects are pneumonitis and lung fibrosis as well as acute cardiovascular collapse.

Preparations

Fluid for injection in ampoules.

SUPPLEMENTARY READING

Drug and therapeutics bulletin 1961; 7: 41.
Gordon PH. Anal fissure – diagnosis and treatment. Consultant 1978: 73.
Jensen SL. Treatment of first episodes of acute anal fissure: prospective randomised study of lignocaine ointment versus hydrocortisone ointment or warm sitz baths plus bran. Br Med J 1986; 292: 1167.
Jackson R. Side effects of potential topical corticoids. Can Med Assoc J 1978; 118: 173.
Rotenberg GN, ed. Compedium of pharmaceuticals and specialities. 12th ed. Toronto: Canadian Pharmaceutical Association, 1977.
Steinberg H. Applied pharmacology related to the colon in diseases of the colon and anorectum. Philadelphia: W.B. Saunders, 1969.

Index

O

Obstetric trauma   129
Ointments   158
Operative trauma   129
Outpatient anal surgery   153

P

Paccini bodies   13
Paget's perianal disease   100
Pain   20
   relief   154
Palpation   27
Papillae   49
Parasitic infection,
   pruritus ani   126
   sexually transmitted   76
Parasympathetic innervation   9
Park's ligament   5
Park's speculum   25
Park's submucous hemorrhoidectomy   47
Pediculosis pubis   126
Penicillin   77
Perianal abscesses   56
   incision and draining   58
   incision and primary suture   58
Perianal furunculosis   91
Perianal hematoma   47
Perianal skin   1
Perianal space   6
Pharmacology   158*ff*
Physiology   11
Pilonidal abscess, acute   68
Pilonidal cyst   68
   chronic   71
Pilonidal sinus   68
   chronic   71
Podophyllin resin   87, 162
Polyp   49
Positioning of the patient
   knee-elbow position   24
   lateral position   23
   lithotomy position   23
Postanal region   7
Postanal repair   132
Premalignant tumors   98*ff*
   Bowen's disease   101
   carcinoma *in situ*   98
   dysplasia   98
   intraepithelial carcinoma   98
   keratoacanthoma   99
   leukoplathia   99
   Paget's perianal disease   100
Pressure profile   15
Primary syphilis   78
Prioderm®   126

Probenecid   77
Proctalgia fugax   120
Proctitis   19
Proctosedyl®   159
Proctosigmoidoscopy   29
Prokainpenicillin   80
Prolapse symptoms   21
Prostate   28
Pruritus ani   121*ff*
Pseudomonas aeruginosa   140
Psoriasis, and pruritus ani   122
Psyllium   161
Pthirus pubis   83
Pudendal nerve   10
Punch biopsy   34

Q

Quadrant sphincterotomy   142

R

Radiation therapy   111
Radiological investigation   31
   Wangensteen-Rice invertogram   37
Rectal,
   capacity   11
   compliance   16
   examination   27
   prolapse   130
   sensation   15
Rectoscopy   29
Rectosigmoidectomy, transanal   136
Rectosphincteric reflex   16
Rectum, resection   109
Reflex, anal   16
Regional anesthesia   151
Resection, abdominoperineal   109
Retardin®   123
Retrovirus   76, 83
Rigid proctosigmoidoscopy   24
Roughage   160
Rubber band ligation   42

S

S plasty   141
Sacral anesthesia   151
Sarcoptes scabiei   83
Sarcoma   117
Scanning
   CT   31
   NMR   31
   ultrasound   32, 58
Schistosoma haematobium   96
Schistosoma japonicum   96
Schistosoma mansoni   96